Handbook of Clinical Electrophysiology of Vision

Minzhong Yu • Donnell Creel
Alessandro Iannaccone
Editors

Handbook of Clinical Electrophysiology of Vision

Springer

Editors
Minzhong Yu
Department of Ophthalmology
University Hospitals Eye Institute
Cleveland, OH
USA

Donnell Creel
Moran Eye Center
University of Utah School of Medicine
Salt Lake City, UT
USA

Alessandro Iannaccone
Center for Retinal Degenerations and Ophthalmic Genetic Diseases
Duke University School of Medicine
Duke Eye Center, Department of Ophthalmology
Durham, NC
USA

ISBN 978-3-030-30416-4 ISBN 978-3-030-30417-1 (eBook)
https://doi.org/10.1007/978-3-030-30417-1

This Springer imprint is published by the registered company Springer Nature Switzerland AG
The registered company address is: Gewerbestrasse 11, 6330 Cham, Switzerland

Preface

Over the past decade, progress in clinical electrophysiology of vision has revealed better understanding of the relationship of retinal pathophysiology and the change of the parameters of electrophysiologic responses and the underlying mechanisms. This handbook, which includes recent updates of techniques, research, and data from clinical electrophysiology of vision, was written by ophthalmic electrophysiologists/vision scientists, and ophthalmologists. We anticipate that clinical researchers and vision scientists will be inspired by developments toward the improvement of diagnoses and drug screening.

Chapters 1, 2, and 3 illustrate the physiologic sources of electrophysiologic responses, the techniques of the recording, and the analysis of electrophysiologic signals. Chapters 4, 5, 6, 7, 8, 9, 10, and 11 summarize the characteristics of the electrophysiologic signals in a number of disorders of the retina and optic nerve, which will help clinicians select the suitable electrophysiology tests for differentiation and diagnosis of diseases.

We would like to thank Springer for giving us the opportunity to publish this book. We are also deeply grateful to all of the chapter authors for their extraordinary contributions. Finally, we thank Xiaoshan Yu, PharmD, RPh, and Feras Mohder, MD, Board-Certified Internist, for their critical review of Chapters 9 and 10, respectively.

Cleveland, OH, USA — Minzhong Yu, PhD, MMed
Salt Lake City, UT, USA — Donnell Creel, PhD
Durham, NC, USA — Alessandro Iannaccone, MD, MS, FARVO

Contents

Contributors

Donnell Creel, PhD Moran Eye Center, University of Utah School of Medicine, Salt Lake City, UT, USA

Roberto Gattegna, MD Center for Retinal Degenerations and Ophthalmic Genetic Diseases, Duke University School of Medicine, Duke Eye Center, Department of Ophthalmology, Durham, NC, USA

Retina Service, Israelitic Hospital, Rome, Italy

Alessandro Iannaccone, MD, MS, FARVO Center for Retinal Degenerations and Ophthalmic Genetic Diseases, Duke University School of Medicine, Duke Eye Center, Department of Ophthalmology, Durham, NC, USA

Wajiha Jurdi Kheir, MD Center for Retinal Degenerations and Ophthalmic Genetic Diseases, Duke University School of Medicine, Duke Eye Center, Department of Ophthalmology, Durham, NC, USA

Alessandro Racioppi, BS Center for Retinal Degenerations and Ophthalmic Genetic Diseases, Duke University School of Medicine, Duke Eye Center, Department of Ophthalmology, Durham, NC, USA

University of North Carolina, Chapel Hill, NC, USA

Alfonso Senatore, MD Center for Retinal Degenerations and Ophthalmic Genetic Diseases, Duke University School of Medicine, Duke Eye Center, Department of Ophthalmology, Durham, NC, USA

Minzhong Yu, PhD, MMed Department of Ophthalmology, University Hospitals Eye Institute, Cleveland, OH, USA

Part I
Basic Theory of Electrophysiology of Vision

Chapter 1
Electroretinography

Donnell Creel and Minzhong Yu

Components of ERG

In 1865, Holmgren noticed that there were electrical changes in an amphibian eye when exposed to light [1]. By 1908, three waves, a, b, and c, of ERG had been identified. In 1933, Ragnar Granit performed several studies on cat retinae manipulating levels of anesthetic to isolate different components contributing to the ERG [2]. He identified the origins of ERG components: a-wave originating from retinal receptors, b-wave from mid-retina (bipolar cells), and c-wave from the retinal pigment epithelium. The 1967 Nobel Prize for Physiology and Medicine was awarded to Ragnar Granit for this work.

Granit labeled three components, PI, PII, and PIII, in the order they were extinguished with deepening the levels of anesthesia in cat. When stimulating with a 2-second pulse of light, a cornea-positive component PI increases to a maximum slowly and then decreases slowly without showing an off response. PII is also a cornea-positive component with short latency reflecting the b-wave, which decreases quickly to a lower positive potential and then shows an off component. PIII best resists anesthesia in cat. It is a cornea-negative component that quickly reaches a minimum potential and remains negative until light stimulus ends.

PIII includes a fast negative PIII and a slow negative PIII [3, 4]. The fast PIII, elicited by the onset of the stimulating light, is from extracellular radial current of photoreceptors, which is caused by the closure of cGMP-gated cationic channels when the photoreceptor outer segments receive light [5, 6]. The slow PIII, maintained when the stimulating light is kept on, is generated by the change of

D. Creel (✉)
Moran Eye Center, University of Utah School of Medicine, Salt Lake City, UT, USA
e-mail: donnell.creel@hsc.utah.edu

M. Yu
Department of Ophthalmology, University Hospitals Eye Institute, Cleveland, OH, USA

M. Yu et al. (eds.), *Handbook of Clinical Electrophysiology of Vision*,
https://doi.org/10.1007/978-3-030-30417-1_1

the transmembrane potential of the Müller cells that is initiated by the reduction of extracellular potassium concentration in the photoreceptor layer during light stimulation. PII is mainly associated with bipolar cell layer. The ERG a-wave is the sum of PIII and PII. The b-wave is mainly associated with PII. Finally, PI is from retinal epithelium layer. The c-wave's main contribution is from PI.

These studies along with other data added to the confirmation of different origins of ERG components a-, b-, and c-waves (Fig. 1.1). The pattern of these waves depends upon stimulus strength and duration.

The clinical application of electroretinograms (ERGs) began in the 1950s, but the common use of clinical ERGs and visually evoked potentials (VEPs) parallels the availability of averaging computers in the 1960s. Electroretinograms and visually evoked potentials have become more sophisticated paralleling advances in computer technology reaching current sophistication in multifocal ERGs and multifocal VEPs.

A flash of light elicits a biphasic waveform recordable at the cornea is similar to the one illustrated (Fig. 1.2). The two components that are most often measured are the a- and b-waves. The a-wave is the first large negative component, followed by the b-wave that is positive and usually larger in amplitude compared with the a-wave (Fig. 1.2).

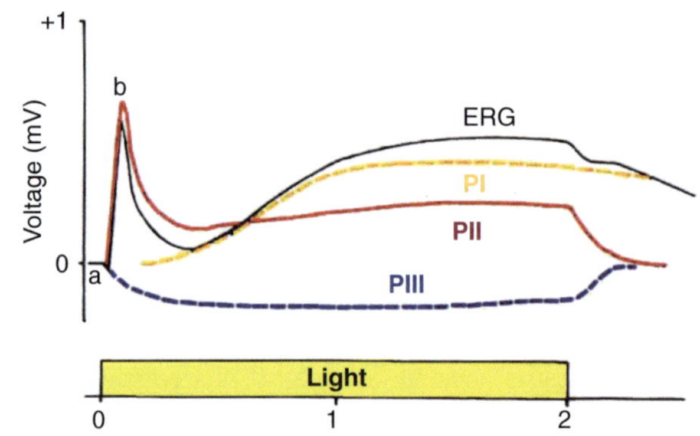

Fig. 1.1 The ERG of a cat following 2 second light stimulus broken down into components affected by depth of ether anesthesia [2]. Components PI, PII, and PIII disappear sequentially as anesthetic depth increased

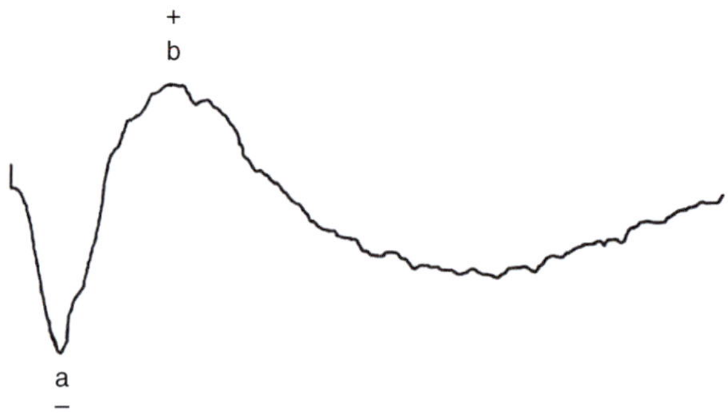

Fig. 1.2 A typical bright white flash ERG waveform

Two principal components of the ERG waveform are quantified in a clinical exam: (1) The amplitude from the baseline to the negative trough of the a-wave and the amplitude of the b-wave measured from the trough of the a-wave to the following peak of the b-wave and (2) the time from flash onset to the trough of the a-wave and the time from flash onset to the peak of the b-wave. These times are referred to as "implicit times" in the jargon of electroretinography. The recording site on the cornea of eye is usually the positive pole of the amplifier and is displayed as upward in ERG; the reference location such as the forehead is the negative pole and displayed downward such as in Fig. 1.2.

The a-wave, also referred to as the "late receptor potential," reflects the overall physiological health of the photoreceptors in the outer retina (Fig. 1.3). By contrast, the b-wave reflects the health of the inner layers of the retina, including the ON bipolar cells and the Müller cells [7], and is affected by the physiology including transmitters of all constituents comprising the mid-retina, horizontal, amacrine, and other types of bipolar cells. Two other waveforms that are sometimes recorded clinically are the c-wave originating in the pigment epithelium [8–11] and the d-wave indicating activity of the OFF bipolar cells [12, 13].

Besides a-, b-, and c-waves, oscillatory potentials (OPs) appear on the ascending limb of b-waves, first described by Cobb and Morton (1954). OPs may be related to

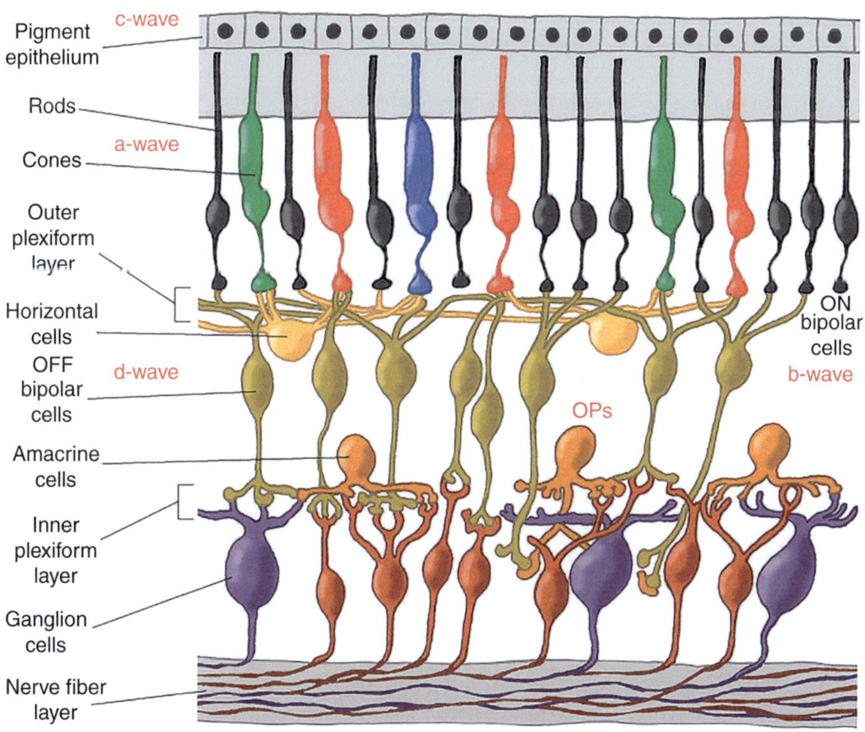

Fig. 1.3 Diagram of retina showing origin of ERG components (Courtesy of Helga Kolb)

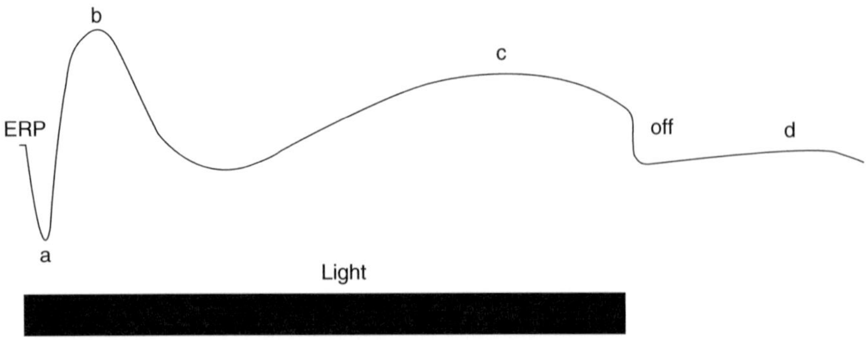

Fig. 1.4 Hypothetical ERG showing all components of the ERG recordable under different conditions

the extracellular currents in the feedback loop among bipolar cells, amacrine cells, and ganglion cells [14, 15].

The early receptor potential (ERP) is a very fast small biphasic component preceding the a-wave, appearing in the first 2 milliseconds following a bright flash, reflecting the earliest chemical responses to light in the receptor outer segments. Approximately 70% of its contribution is from cones. The ERP latency is less than 60 microseconds. Due to photovoltaic effects, ERPs are best recorded using non-metal electrode, such as with a cotton wick. The ERP is difficult to record and is not commonly used clinically. Figure 1.4 depicts a hypothetical ERG showing all components of the ERG recordable under different conditions.

The ERG of a normal full-term infant looks similar to an adult ERG. The ERG amplitude in a newborn infant is occasionally small in the first couple of months. The ERG attains peak amplitude in adolescence and slowly declines in amplitude throughout life [16]. After age 60, the amplitude of the ERG declines even more. Implicit times slow gradually from adolescence through old age as well.

Types of ERG

Full-Field ERG

Full-field ERGs (ffERGs) are recorded with full-field flash stimulation in dark-adapted and light-adapted conditions. The testing sequence and flash stimuli vary between protocols. Combined with colors of flashes and rate of stimulation (i.e., single and flickering flashes), one can separate rod and cone function. The full-field ERG is the preferred test for assessing retinal dysfunction expressed throughout the retina. For patients over 5 years of age, most laboratories use a Ganzfeld (globe) with a chin rest and fixation points (Fig. 1.5). The Ganzfeld

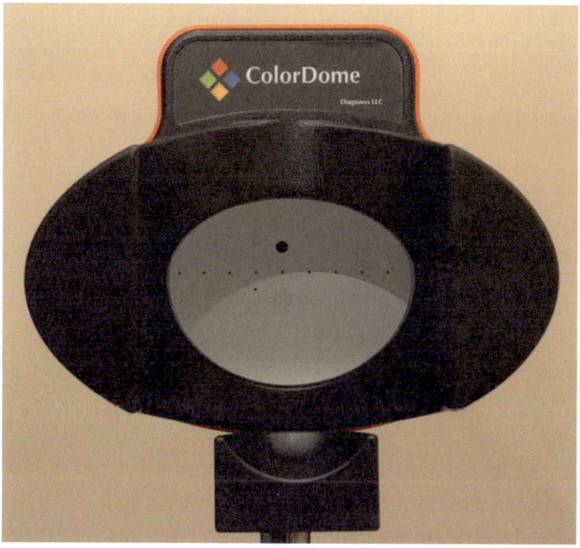

Fig. 1.5 An example of a Ganzfeld

allows the best control of background illumination, stimulus color, and flash intensity. Strobe lamp, LED flash, or Ganzfeld flash stimulation can be used to record the ERG following a single flash or to average responses to several flashes with the aid of a computer. Clinical decisions can be made from ERGs generated by all methods.

Dark-Adapted Rod-Driven Scotopic and Mixed Flash ERGs

Suggested protocols recommend dark adaptation before recording ERGs for a set time, minimally 20 minutes, with some laboratories using 30 minutes or longer. After dark adaptation, dim flashes of light are used to measure rod function. The most common protocol is the ISCEV standard white-flash protocol [17]. To minimize the response of cone photoreceptors during the testing of rod function, a dim blue flash stimulation with peak spectral color near 450 nm producing normal b-wave amplitude of about 200 μV is suggested to use prior to all-white flash ISCEV protocol. In addition, a dim red flash color with wavelength of above 585 nm producing a normal b-wave amplitude of about 200 μV can also be used before starting the all-white flash ISCEV protocol. The scotopic dim red flashes stimulate both rod and cone photoreceptors producing an a-wave followed by an early cone-dominated x component and then a larger slower rod b-wave [18]. The scotopic dim blue and red flashes can be created by placing Kodak Wratten filters 47 and 26 in front of a dim strobe flash or selected on LED stimulus software, such as using photopic blue 0.0004 $cd{\cdot}s{\cdot}m^{-2}$ and photopic red 0.2 $cd{\cdot}s{\cdot}m^{-2}$ (Fig. 1.6).

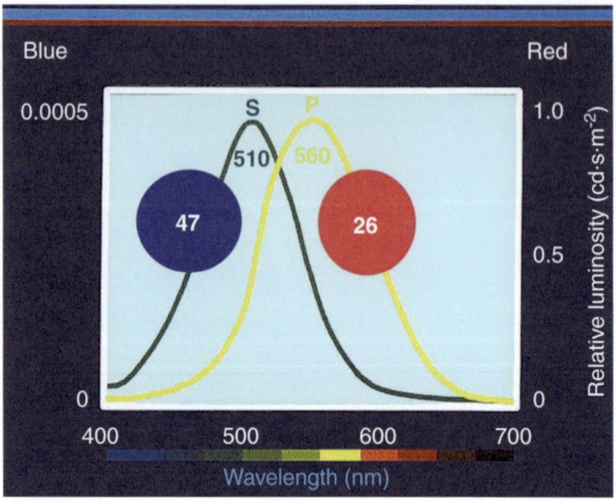

Fig. 1.6 Visual spectrum of 450 nm dim blue and above 580 nm dim red light flashes in relation to sensitivity curves of rods and cones showing example Wratten color filters used at dim flash intensities to isolate rod and cone physiology

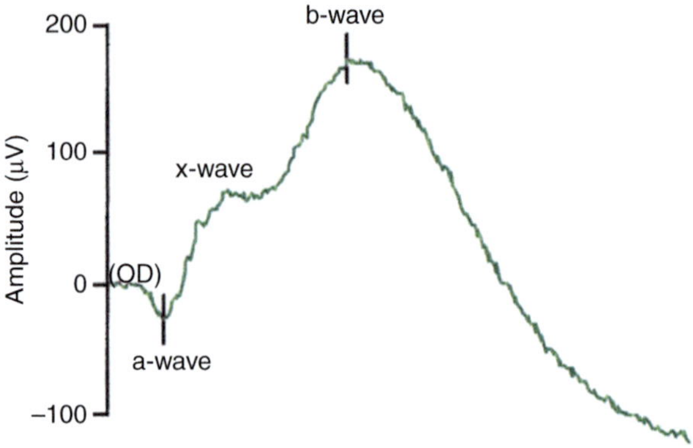

Fig. 1.7 Scotopic dim red flash ERG showing x component

A scotopic dim red flash ERG verifies rod function quantified by b-wave amplitude and produces a fast x component resembling the photopic white flash ERG created by the slight stimulation of cones [18]. The x component reflects cone physiology, which can be considered along with photopic single white flash ERGs and 30 Hz flicker to judge cone function (Fig. 1.7).

Bright white flashes in the dark-adapted condition produce the large-amplitude ERGs characterized by large-amplitude a-wave appearing at short implicit time followed by a large-amplitude b-wave with oscillatory potentials on the ascending limb. These scotopic bright white flash ERGs show mixed rod-cone responses for assessment of rod and cone functions. However, these ERGs are not as sensitive to subtle changes in retinal pathology.

Light-Adapted Cone-Driven ERG

Light-adapted ERG responses are recorded with background illumination of 30 cd·m^{-2} after 10 minutes of light adaptation. Single photopic bright white flash ERGs and 30 Hz flicker are used to assess cone function along with the x component of the scotopic dim red flash ERG.

Pattern ERG

The pattern electroretinogram (PERG) is the retinal response elicited by a pattern reversal stimulus that assesses retinal ganglion cell function. Clinically, PERG is used in patients including those with glaucoma, optic neuropathies, and ganglion cell diseases. Its amplitude is small, usually less than 10 μV (Fig. 1.8), including a small initial negative component with latency of approximately 35 ms (N35), a much larger positive component peaking around 50 ms (P50), and followed by a large negative component around 95 ms. Macular function is reflected in the P50 component and retinal ganglion cell function by the N95 component [19, 20].

For the PERG test, some contact lens electrodes cannot be used, because they blur the pattern stimulation. Thin conductive electrodes such as the Dawson, Trick, and Litzkow (DTL) or Arden gold foil are preferable. Each laboratory should establish its normal values. PERG changes with age, although it is stable from teens through age 55 years. See ISCEV pattern ERG standard for detailed methodology [19].

Multifocal ERG

A limitation of the traditional global, full-field ERG is that the recording is a mass potential reflecting the health of the whole retina, which may not be sensitive to changes in small areas of the retina. Unless 20% or more of the retina is affected

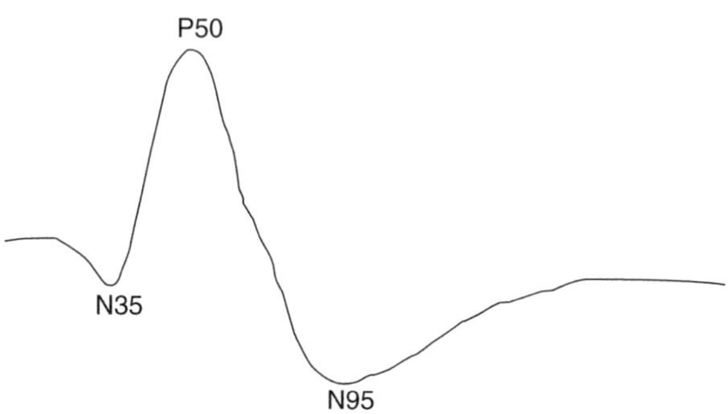

Fig. 1.8 Representative pattern ERG. Time epoch 250 msec

with a diseased state, the full-field ERG is usually normal. A legally blind person with macular degeneration, enlarged blind spot, limited schisis, or other small central scotomas may have a normal full-field ERG. The most important development in ERGs in the past 25 years is the multifocal ERG (mfERG). Erich Sutter adapted the mathematical sequences called binary m-sequences, creating a software that can extract hundreds of focal ERGs (fERG) from a single channel of electrical signal. This system allows assessment of ERG activity in many small areas of retina. With this method, the ERG responses from hundreds of retinal areas can be recorded in a short time [21]. Conventional ERG electrodes can be used to record mfERGs from the cornea of a dilated eye. The first kernel and the second kernel of mfERG waveforms are derived. The first kernel reflects the response to the current stimulation. The second kernel reflects the current response with the interaction of a previous response. The second kernel first slice reflects the current response with the interaction of the closest previous response. The second kernel second slice reflects the current response with the interaction of the previous response with one more interval. The waveform of mfERG of first kernel and second-order kernel first slice in each trace mainly consists of an initial negative component N1 followed by a positive peak P1, subsequent N2, and even P2 components [22, 23]. Although the first negative component N1 resembles a-wave and P1 resembles b-wave in full-field ERG, they do not reflect the responses from the exact same retinal cell layers. mfERGs mainly reflect the responses of on- and off-bipolar cells with small contributions from photoreceptors and inner retina [24]. Therefore, mfERGs mainly reflect the status of retinal layers of bipolar cells and photoreceptors because the response of bipolar cells is also affected by the input from photoreceptors. Besides the analysis in different layers of retina, mfERGs mainly reflect the status of different areas of retina. Small scotomas in retina can be mapped, allowing the degree of retinal dysfunction to be quantified. Most mfERG analyses are based on amplitudes and implicit times of the mfERG components. See Hood et al. (2012) for recommended ISCEV International Standard mfERG protocol [25]. A normal mfERG array is displayed in Fig. 1.9.

The mfERG is the test of choice to map central scotomas of any etiology originating in the retina in the central 60° of visual fields. The common mfERG examination, due to bright stimulus field, only reflects physiology of cone pathways. Multifocal ERGs are not the test of choice for generalized retinal dysfunction. For differentiating between cone, cone–rod dystrophies, rod–cone dystrophies, and related disorders, it is better to record full-field ERGs.

Focal Flash ERG

The focal ERG (fERG) is used to assess function of a single small area of the retina, often in the foveal, macular area. Stimulus field sizes from 3° to 18°, and different stimulus frequencies are used. Focal ERGs along with mfERGs are useful to quantitate

Fig. 1.9 Normal multifocal electroretinograms (mfERGs)

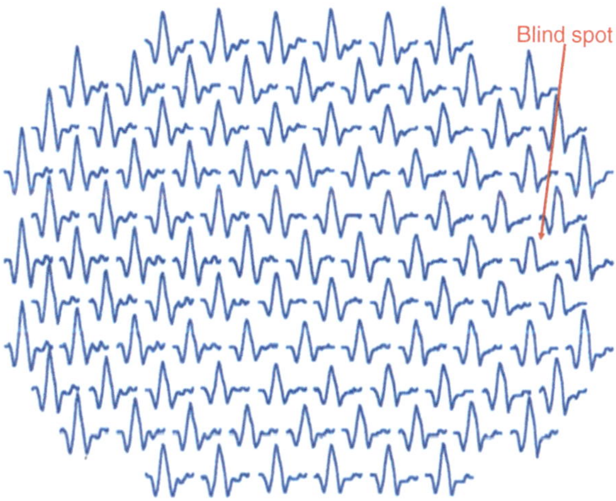

macular function including in patients with age-related macular degeneration (AMD). Multifocal ERGs produce a map of up to 60° of central visual field, whereas fERG assesses only one area. Both mfERGs and fERG require good fixation.

Electrodes, Stimulation, and Recording Methods

ERG Electrodes

Figure 1.10 shows common ERG electrodes.

The Burian–Allen contact lens electrode is usually bipolar so that another negative reference electrode is not needed. It includes a speculum coated with silver that holds the eyelids open and functions as the negative reference electrode. The central contact lens with a wire ring "floats" on the cornea supported by a small spring as the positive active electrode. These electrodes are available in different sizes that fit the eye from full-term newborns to adults.

ERG–Jet electrode is a type of monopolar electrode that requires a negative reference location such as forehead, temporal area next to the canthus, or ear lobe.

The Dawson, Trick, and Litzkow (DTL) disposable electrode, a thin silver thread inserted at lower lid margin, is popular. They are monopolar electrodes. Although the signals recorded by DTL is lower than contact lens electrode, it is recommended for the eyes with damaged cornea, including the eyes undergoing LASIK surgery, corneal transplantation, and corneal abrasion, or for recording of PERG that requires the patient to see the stimulating pattern as clear as possible, and can be used in general. In addition, local anesthesia is not always required for using DTL electrode and are more comfortable to the patient, which are the other advantages of this electrode.

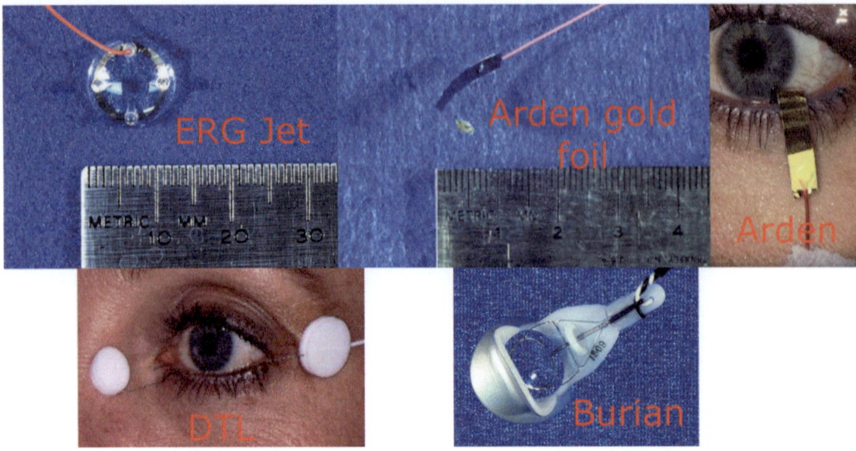

Fig. 1.10 Commonly used ERG recording electrodes

The ERG can also be recorded using skin electrodes placed near the eye, such as the sensor strips of LKC Technologies. This skin electrode set includes active, reference, and ground electrodes in the strip, which can be used with either a conventional electrophysiologic device or with an LKC RetEval system that is quite portable and not limited to having an electrical outlet. Since skin electrodes are not in direct contact with the eye, the recorded ERG amplitude is significantly lower.

If electrodes are to be reused, then they should be sterilized with a solution that neutralizes prion-transmitted diseases such as Creutzfeldt–Jakob disease (CJD). Follow sterilization recommended by the manufacturer to avoid damage to the electrode.

Stimulation

There are several methods of stimulating the eye. Some laboratories use a strobe lamp, array of LEDs, or hand-held LED stimulators that are mobile and can be easily placed in front of a patient sitting or reclining (e.g., LKC RetEval, DiagnosysLLC Colorburst). The mobility of these stimulators is preferable in situations such as at the hospital bedside or in the operating room.

For eliciting pure rod response, low flash luminance ISCEV dark-adapted 0.01 cd·s·m^{-2} flash is commonly used. In addition, a lower intensity blue flash can also be used, which is most sensitive not only to rod disorders but also to systemic metabolic aberrations and retinal toxicity. B-wave amplitudes elicited by scotopic dim blue flash and red flashes can be adjusted in flash intensity so that similar b-wave amplitudes are obtained with each stimulus. Examples include near 0.0004 cd·s·m^{-2} photopic blue flash intensity and 0.2 cd·s·m^{-2} photopic red flash intensity. This matching is called "scotopically balanced" blue and red flashes [18]. In clinical application, the b-wave amplitudes elicited by dim "scotopically balanced" blue and red flashes are most sensitive to detecting changes in some disorders.

Rods are about three log units more sensitive than cones. However, cones recover faster than rods. Using different rates of stimulus presentation, flicker, plus levels of flash intensity allow the rod and cone contributions to the ERG to be separated. Even under ideal conditions, rods cannot follow a flickering light up to 20 per second, whereas cones can easily follow a 30 Hz flicker. A rate of 30 Hz is routinely used to test if a retina has good cone physiology.

Recording Method

The recommended ISCEV standard ERG series includes six protocols. See McCulloch et al. (2015) for details [17, 26].

1. Dark-adapted 0.01 cd·s·m^{-2} ERG
2. Dark-adapted 3 cd·s·m^{-2} ERG
3. Dark-adapted 10 cd·s·m^{-2} ERG
4. Dark-adapted oscillatory potentials
5. Light-adapted 3 cd·s·m^{-2} ERG
6. Light-adapted 30 Hz flicker ERG

About 20% of ERG laboratories add scotopic dim blue flash and/or scotopic dim red flash prior to the first dark-adapted 0.01 cd·s·m^{-2} ERG. Example flash intensities for the scotopic dim blue flash intensity are photopic 0.0004 cd·s·m^{-2} for 440 nm blue, and for dim red flash photopic 0.2 cd·s·m^{-2} for 600–635 nm red.

Dilation, Dark Adaptation, and Topical Anesthesia

The ERG can be recorded in several ways. The pupil is usually dilated with drops of 1% tropicamide and 2.5% phenylephrine HCl for adults or 1% cyclopentolate HCl for children. The patient should dark adapt for 20–30 minutes. A total of 30 minutes or more in the dark produces a state of 98% dark adaptation in most individuals. The corneal surface is usually anesthetized with 0.5% proparacaine HCl.

Testing Infants

Electrophysiology can be more helpful with pediatric patients as children are less aware and unlikely to notice visual dysfunction or changes in vision that an adult would notice, such as a visual field change, loss of acuity, and abnormal color vision. A child will usually accept changes in vision and accept how they differ from others, even blindness in the very young, as normal.

Infants up to 2 years of age can usually be tested without sedation with the parent holding them bundled in a blanket. It is difficult to convince a child less than 5 years of age to allow a contact lens or speculum electrode in their eye. Silver wire or gold foil usually picks up too much muscle artifact from an active child. Skin electrodes below the eye are an option. Alternatively, ophthalmic electrophysiology can also be recorded as part of exams under anesthesia (EUA). The child can be sedated or anesthetized in sedation unit or operating room. Oral glucose or sucrose are part of the options for analgesia during the procedures of ERG recording in infants up to 18 months of age [27]. Combining remifentanil and propofol can also be used for the recording of ERG and visually evoked potential (VEP) under anesthesia.

When recording ERGs or VEPs under anesthesia, depth of anesthesia must be considered. Surgical-depth anesthesia interferes with recording electrophysiology, especially visual evoked potentials. Anesthesia affects the ERG and VEP in different degrees with different types and depths of anesthesia. Lighter level of anesthesia has less effect. Some anesthetics can attenuate b-wave amplitude as much as 50%. Most anesthetics do not affect a-wave amplitude or implicit time. One must coordinate with the anesthesiologist to attain a light level of anesthesia. When recording VEPs, the patient's EEG can show whether the patient is lightly anesthetized. If EEG amplitude is low, then ask the anesthesiologist to lighten anesthetic depth and wait a few minutes.

Most laboratories do not have a Ganzfeld stimulator that can be placed over the face of a sedated patient in the operating room. Thus, flash stimulations are often delivered with a handheld strobe or LED stimulator placed over eye. It is difficult to completely darken the operation room; thus, abbreviated testing is more easily accomplished under mesopic and photopic light conditions using mesopic single flashes, oscillatory potentials, and 30 Hz flicker.

Advanced Analysis of ERG Data

Intensity-Response Function of ERG B-Wave

In scotopic and photopic ffERG, the b-wave amplitude generally increases with flash luminance. An intensity–response (I-R) curve can be plotted showing the change of b-wave amplitude with the increase of the flash luminance (Fig. 1.11). When the flash luminous intensity is low, the b-wave amplitude increases slowly with the increase of the flash luminance. After reaching a certain level

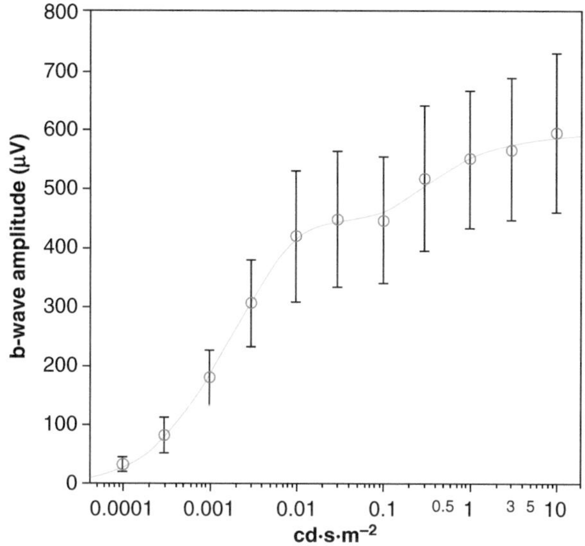

Fig. 1.11 Intensity-response function of ERG b-wave showing the increase of b-wave amplitude with the increase of flash intensity

of luminance, the b-wave amplitude increases faster with the increasing luminous intensity. Then, increasing ERG amplitude gradually saturates at a high flash luminance. After saturation, the b-wave amplitude may decrease mildly with the increase of the flash luminous intensity. As luminance increases, the I–R curve of the b-wave amplitude shows a second limb of b-wave amplitude in different slope. In scotopic ERG, the second limb of the I–R curve ascends with a different slope associated with the additional cone contribution on the rod response. In photopic ERGs, the second limb of the I–R curve descends [28, 29], which is related to the balance of the ON- and OFF-hyperpolarizing bipolar cells in different flash luminous intensity [30].

For quantitatively analysis of the I–R curve in both scotopic and photopic ERGs, the first limb, a sigmoid growth curve, can be fit by Naka–Rushton equation that was derived based on the research of Naka and Rushton in 1966 [31]:

$$R = R_{\max} \times I^n / \left(I^n + K^n\right),$$

where R is the b-wave amplitude, $R_{\max}$ is the maximum of the first limb of the b-wave amplitude, I is the flash luminance, n is the slope of the curve, and K is the flash luminance which elicits half of the maximal b-wave amplitude.

Oscillatory Potentials (OPs) Analysis

The frequency lower limits for recording OPs components are from 75 to 100 Hz, and the upper limits are 300 to 1000 Hz. It is more convenient to analyze the OPs with digital filtering and frequency spectrum analysis in which a Fast Fourier

Transform (FFT) is used. During this process, the frequency spectrum is calculated from the raw ERG waveform, in which the amplitudes of the components in different frequencies are derived. The frequency spectrum can be used directly for the comparison with that of the control group [32]. Another method is to use band-pass filtering, selecting the components of the frequency spectrum in the range of about 100–150 Hz to extract the OPs waveform. Then, the amplitudes of the several OPs wavelets can be analyzed, respectively.

References

1. Holmgren F. Metod att objektivera effektenav ljusintryck pa retina. Upsala lakaref Forhandl. 1865;1:177–91.
2. Granit R. The components of the retinal action potential in mammals and their relation to the discharge in the optic nerve. J Physiol. 1933;77(3):207–39.
3. Murakami M, Kaneko A. Subcomponents of P3 in cold-blooded vertebrate retinae. Nature. 1966;210(5031):103–4.
4. Sillman AJ, Ito H, Tomita T. Studies on the mass receptor potential of the isolated frog retina. I. General properties of the response. Vision Res. 1969;9(12):1435–42.
5. Penn RD, Hagins WA. Signal transmission along retinal rods and the origin of the electroretinographic a-wave. Nature. 1969;223(5202):201–4.
6. Sillman AJ, Ito H, Tomita T. Studies on the mass receptor potential of the isolated frog retina. II. On the basis of the ionic mechanism. Vision Res. 1969;9(12):1443–51.
7. Miller RF, Dowling JE. Intracellular responses of the Muller (glial) cells of mudpuppy retina: their relation to b-wave of the electroretinogram. J Neurophysiol. 1970;33(3):323–41.
8. Taumer R, et al. Experiments concerning the human C-wave. Albrecht Von Graefes Arch Klin Exp Ophthalmol. 1976;198(2):139–53.
9. Yu M, Peachey NS. Use of direct current electroretinography for analysis of retinal pigment epithelium function in mouse models. Methods Mol Biol. 2018;1753:103–13.
10. Skoog KO, Nilsson SE. Changes in the c-wave of the electroretinogram and in the standing potential of the eye after small doses of toluene and styrene. Acta Ophthalmol. 1981;59(1):71–9.
11. Wu L, Lurie M, Marmor MF. The C-wave of the rabbit electroretinogram during dark-adaptation and the steady-state. Acta Ophthalmol. 1981;59(4):603–8.
12. Ueno S, et al. Contribution of retinal neurons to d-wave of primate photopic electroretinograms. Vision Res. 2006;46(5):658–64.
13. Naarendorp F, Williams GE. The d-wave of the rod electroretinogram of rat originates in the cone pathway. Vis Neurosci. 1999;16(1):91–105.
14. Wachtmeister L, Dowling JE. The oscillatory potentials of the mudpuppy retina. Invest Ophthalmol Vis Sci. 1978;17(12):1176–88.
15. Gouras P, MacKay C. A new component in the a-wave of the human cone electroretinogram. Doc Ophthalmol. 2000;101(1):19–24.
16. Weleber RG. The effect of age on human cone and rod ganzfeld electroretinograms. Invest Ophthalmol Vis Sci. 1981;20(3):392–9.
17. McCulloch DL, et al. ISCEV Standard for full-field clinical electroretinography (2015 update). Doc Ophthalmol. 2015;130(1):1–12.
18. Thompson DA, et al. ISCEV extended protocol for the dark-adapted red flash ERG. Doc Ophthalmol. 2018;136(3):191–7.
19. Holder GE, et al. ISCEV standard for clinical pattern electroretinography – 2007 update. Doc Ophthalmol. 2007;114(3):111–6.

20. Bach M, et al. ISCEV standard for clinical pattern electroretinography (PERG): 2012 update. Doc Ophthalmol. 2013;126(1):1–7.
21. Sutter EE. Noninvasive testing methods: multifocal electrophysiology. In: Dartt DA, editor. Encyclopedia of the eye. Oxford: Academic Press; 2010.
22. Luu CD, et al. Multifocal electroretinogram in children on atropine treatment for myopia. Br J Ophthalmol. 2005;89(2):151–3.
23. Klemp K, et al. The multifocal ERG in diabetic patients without retinopathy during euglycemic clamping. Invest Ophthalmol Vis Sci. 2005;46(7):2620–6.
24. Hood DC, et al. Retinal origins of the primate multifocal ERG: implications for the human response. Invest Ophthalmol Vis Sci. 2002;43(5):1673–85.
25. Hood DC, et al. ISCEV standard for clinical multifocal electroretinography (mfERG) (2011 edition). Doc Ophthalmol. 2012;124(1):1–13.
26. Marmor MF, et al. ISCEV Standard for full-field clinical electroretinography (2008 update). Doc Ophthalmol. 2009;118(1):69–77.
27. Pasek TA, Huber JM. Hospitalized infants who hurt: a sweet solution with oral sucrose. Crit Care Nurse. 2012;32(1):61–9.
28. Rufiange M, et al. The photopic ERG luminance-response function (photopic hill): method of analysis and clinical application. Vision Res. 2003;43(12):1405–12.
29. Rufiange M, et al. Cone-dominated ERG luminance-response function: the Photopic Hill revisited. Doc Ophthalmol. 2002;104(3):231–48.
30. Sieving PA, Murayama K, Naarendorp F. Push-pull model of the primate photopic electroretinogram: a role for hyperpolarizing neurons in shaping the b-wave. Vis Neurosci. 1994;11(3):519–32.
31. Naka KI, Rushton WA. S-potentials from colour units in the retina of fish (Cyprinidae). J Physiol. 1966;185(3):536–55.
32. Yu M, Peachey NS. Attenuation of oscillatory potentials in nob2 mice. Doc Ophthalmol. 2007;115(3):173–86.

Chapter 2
Visually Evoked Potentials

Donnell Creel and Minzhong Yu

Hans Berger made the first electroencephalogram (EEG) recording on July 6, 1924, during a neurosurgical operation performed by the neurosurgeon Nikolai Guleke. Berger reported the EEG in 1929, also introducing the terms alpha and beta waves. During the 1930s, clinical EEGs were recorded in the USA and Canada. When a flash of light is presented, single visually evoked potential can often be seen in the EEG recorded from occipital scalp.

Evoked potentials, whether auditory, somatosensory, or visual, are extracted from the EEG by a simple program. This technique of extracting a signal from noise is one of the oldest applications of computer technology. This process is similar to the programs used to extract radar signals from jamming during World War II. Adding the electrical activity for a time period is called "signal averaging". Dawson first demonstrated a signal-averaging device in 1951. Signal-averaging computers have been commercially available since the early 1960s. The computer programs save a defined time period of EEG activity following a sensory stimulus that is repeated over and over, adding the signals together. The random EEG activity averages away, leaving the sensory evoked potential. Depending on the signal-to-noise ratio, an averaged evoked potential can be visible following only a few stimuli.

The terms visually evoked potential (VEP), visually evoked response (VER), and visually evoked cortical potential (VECP) are equivalent. They refer to electrical potentials, initiated by visual stimuli that are recorded from the scalp overlying visual cortex. VEPs are used primarily to measure the functional integrity of the visual pathway from retina to the visual cortex. VEPs are better at quantifying the functional integrity of the visual pathway than scanning techniques such as magnetic resonance imaging (MRI).

D. Creel (✉)
Moran Eye Center, University of Utah School of Medicine, Salt Lake City, UT, USA
e-mail: donnell.creel@hsc.utah.edu

M. Yu
Department of Ophthalmology, University Hospitals Eye Institute, Cleveland, OH, USA

M. Yu et al. (eds.), *Handbook of Clinical Electrophysiology of Vision*,
https://doi.org/10.1007/978-3-030-30417-1_2

Any abnormality that affects the visual pathway or visual cortex in the brain can affect the VEP. Examples are cortical blindness due to meningitis or anoxia, optic neuritis as a consequence of demyelination, optic atrophy, stroke, and compression of the optic pathways by tumors, amblyopia, neurofibromatosis, and retinal, especially macular, dysfunction. Generally, demyelination of the optic nerve slows the speed of neural transmission of the VEP, and compression of the optic nerve reduces the amplitude of VEPs. Macular damage can also slow the VEP latency and reduce its amplitude.

Origins of VEP Components

A representative pattern-reversal VEP waveform is depicted in Fig. 2.1. The pattern-reversal VEP produces the most reliable waveform between individuals. A definitive negative N1 component appears near 75 ms, a prominent positive P1 component peaks near 100 ms, followed by a negative N2 component at about 135–145 ms with variable time. These times depend on the size of pattern, brightness, sharpness, and speed of the pattern change.

Contributions to VEP waveform recorded from mid-occipital scalp are affected by sources throughout the posterior third of the brain. Youssofzadeh et al. (2015) demonstrated the most active areas of the brain at different time points following pattern stimulation according to the maximal activity of EEG and functional MRIs [1]. As depicted in the functional visual fMRIs in Fig. 2.2, the brain activity during pattern stimulation varies considerably in the occipital area, depicting the highest activity in occipital visual areas while the subject viewed a pattern. Note that there is high interhemispheric variation. In this test, the visual stimulus pattern subtended a 30-degree field, displaying a checkerboard pattern with 100% contrast and reversal rate of four times per second. During visual stimulation, a number of dipole fields are generated, combining in an additive fashion to create the surface VEP [2].

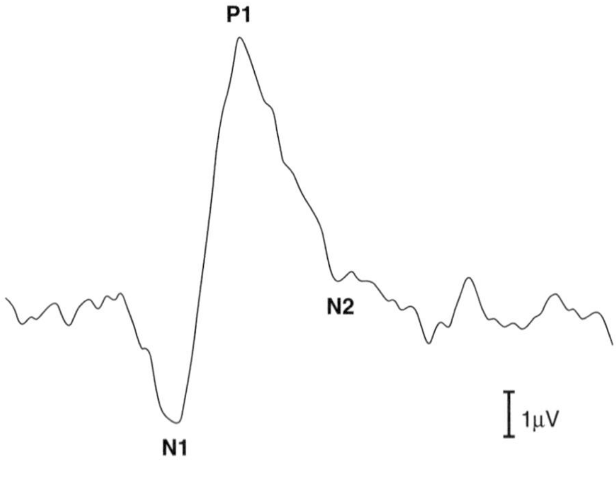

Fig. 2.1 Typical normal pattern-reversal VEP waveform. Time epoch 250 msec

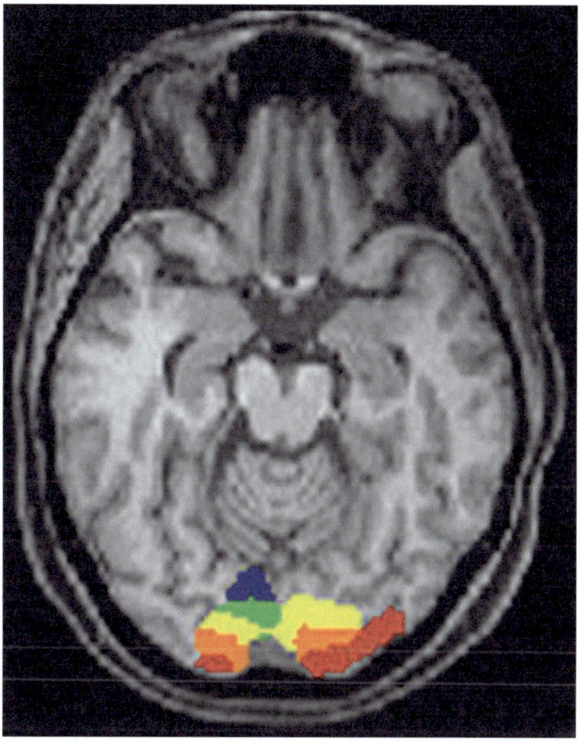

Fig. 2.2 Functional MRI displays highest activity in occipital visual areas, while the subject viewed checked pattern stimulation showing asymmetry

The neural generators of the VEP components are not clearly defined. Research with multichannel scalp recordings, visual MRI activity, and dipole modeling supports the interpretation that the visual cortex is the source of the early component of the VEP N1 [3]. The early phase of the P1 component with a peak around 95–110 ms is likely generated in the dorsal extrastriate cortex of the middle occipital gyrus. The later negative component N2 is generated from several areas including a deep source in the parietal lobe [4].

The distribution of VEP activity varies considerably in the occipital area. A number of dipole fields are generated resulting in a complicated interaction [2]. These multiple sites of generators interact at different levels in the visual areas, making the source localization difficult when making individual clinical decisions. Because of individual idiosyncrasies in occipital anatomy and visual projections, one cannot make the assumptions about the distinguishing of the VEP sources as clearly as those in electroretinograms and auditory brainstem responses.

Although fine-tuning of the maturation of the visual system contributing to improving acuity continues well into school-age children, anatomical studies of foveal cone packing indicate that the retina of an infant of 6 months is already capable of 20/20 vision, while preferential looking studies indicate 20/60 visual acuity by 6 months of age. As the optic pathways add myelin from birth through childhood, the peak time of the P1 component decreases from as long as over 200 ms at birth to near 100 ms of an adult by age 7 years. Retinal development, cortical cell density, and myelination are close to that of an adult by age 6 years. Children by this age produce adult VEP waveforms. The two periods of life with greatest variation in VEP peak times and waveform within an age range are preschool years, birth to age 6, and after age 55.

Electrode Placement

VEPs can be recorded from many scalp locations in humans. Visual stimuli stimulate both the primary visual cortex and the secondary areas of the cortex. Clinical VEPs are usually recorded from mid-occipital scalp overlying the calcarine fissure. This location is the closest to primary visual cortex (Brodmann's area 17). A common system for placing electrodes is the "10–20 International System" that is based on the measurements of head size [5]. The mid-occipital electrode location (Oz) is the active electrode on the midline (Fig. 2.3). The Oz location is calculated as 10% of the distance between the inion and nasion, which is 3–4 cm above the inion in most adults. Lateral occipital electrodes are at the similar distance (i.e., 3–4 cm) off the midline at the level on the inion. The negative reference electrode is usually put on the forehead at the midline. The ground electrode is usually on an earlobe or temple. Electroencephalographic skin electrodes are used for the active (positive), reference (negative), and ground electrodes. Electrode gel or paste should be included between the skin and the electrode after thoroughly cleaning of the skin to reduce the electrical resistance. Some laboratories, and unique applications, have other preferred scalp locations. See Odom et al. (2016) for ISCEV standard for VEP recording [6].

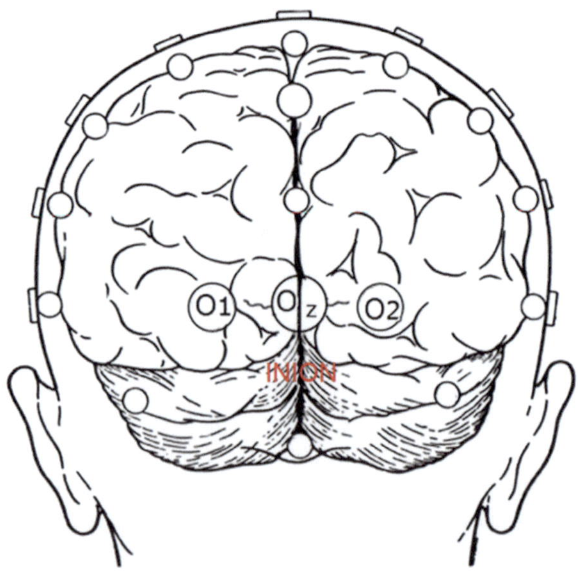

Fig. 2.3 The 10–20 system electrode placement for occipital scalp VEP recording

Amplifier and Analysis

Direct Current (DC)-coupled or Alternating Current (AC)-coupled amplifiers with at least 10 MΩ of input impedance in the frequency range of 50–60 Hz can be used. The bandpass of the amplifier should include 1–100 Hz. The sampling rate of VEP should be at least 1000 Hz. Notch filter of 50 or 60 Hz, which may distort VEP waveform, is not recommended if the interference does not affect the recording and the averaged waveform is acceptable. If the interference is too high, then it may be helpful to rearrange the locations of the preamplifier and the patient. Many VEP waveforms are averaged to obtain the mean VEP waveform. The amplitude of the noise in the averaged waveform reduces as the square root of the number of averaged samples. The number required to obtain the averaged VEP waveform depends on the signal-to-noise ratio of the original signal, which is at least 50 in a clinical setting. However, it takes longer to average more waveforms, which may cause fatigue of the patient. Inattention and muscle artifact may cause variation of the VEP waveforms. At least two averaged VEP waveforms in each condition should be obtained to verify the reproducibility. If the two averaged VEP waveforms are not similar, the test should be repeated. The sweep duration should be from 250 to 500 ms from the onset of the stimulation, according to the implicit times of the VEP components in specific conditions. The amplifier must be electrically isolated from the patient according to the safety standards of medical recording systems.

Types of VEP

Steady-State and Transient VEP

According to the stimulating frequency, VEPs are classified as steady-state and transient VEPs. Steady-state VEPs are those recorded using stimulation rates of more than 3 Hz, in which the next VEP response starts before the previous response finishes. Transient VEPs are recorded using the stimulating rates of less than 3 Hz, producing components of different latencies that can be followed during maturation, pathological conditions, and changes of visual acuity.

Flash and Pattern VEP

Depending on the type of stimulation, VEPs are classified as flash or pattern VEPs. Flash VEPs are used in patients who cannot fixate on pattern stimulation or with cloudy cornea, lens, or vitreous body. It has three positive peaks and three negative peaks (Fig. 2.4). The disadvantage of flash VEPs is that the intrasubject and intersubject variation of the implicit times and amplitudes of the VEP components are larger. For example, the implicit time of the P2 (P110) component of flash VEP has intersubject standard deviation of about 10 ms.

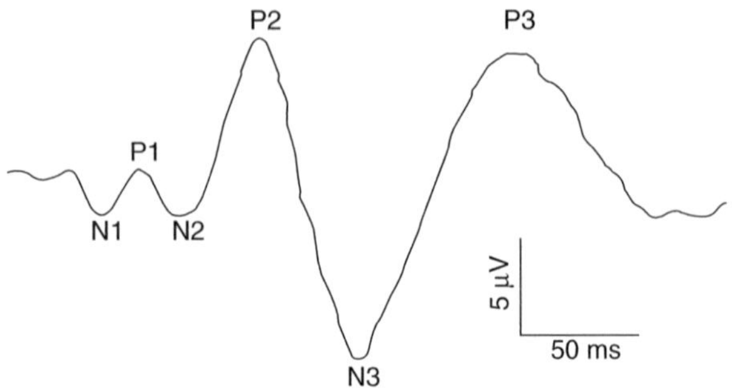

Fig. 2.4 Typical flash VEP waveform

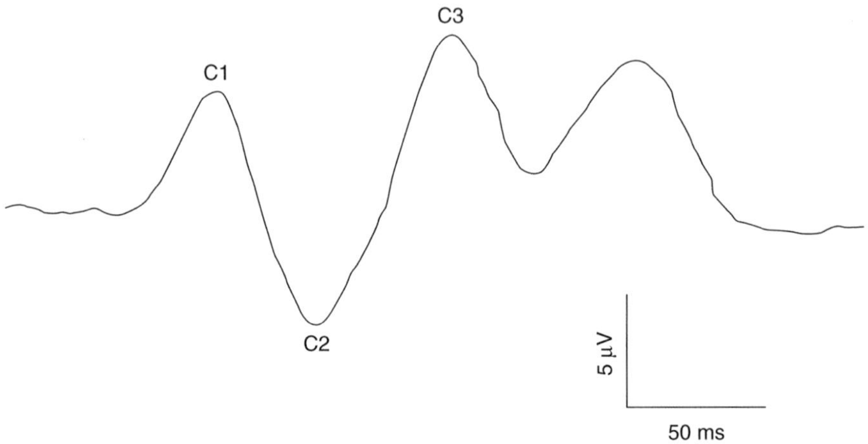

Fig. 2.5 Typical pattern-onset VEP waveform

Pattern VEPs are elicited by pattern onset/offset or pattern reversal of the checkerboard or grating pattern. In pattern onset stimulation, the pattern is made to appear (onset) and disappear (offset). The pattern onset VEP usually includes a positive component C1 at about 80 ms and a large negative component C2 at about 110 ms, opposite to those of flash and pattern-reversal VEPs (Fig. 2.5). Pattern onset stimulus produces a more robust VEP from the central 30° of retina than pattern-reversal stimulus [7]. Pattern onset stimuli can also be the preferable stimulus in conditions such as monitoring amblyopia, in patients with poor acuity, possible hemispheric asymmetry, or a defect in geniculo-striate projections. Pattern onset stimuli are also useful in examining patients that might be malingering because pattern onset is less sensitive to poor fixation. While pattern onset VEPs has lower intrasubject variation, it has higher intersubject variation. For example, the C2 (N110) component of the pattern onset VEP has intersubject standard deviation of about 10 ms. Pattern-reversal VEP has smaller standard deviation (about 6 ms) for the P1 component (Fig. 2.1). However, pattern-reversal stimuli usually exacerbate nystagmus in some patients that interferes with fixation during the VEP test. Pattern-reversal stimuli should not be used in patients with nystagmus.

When recording ERGs or VEPs under anesthesia, depth of anesthesia must be light. Surgical-depth anesthesia interferes with recording electrophysiology, especially visually evoked potentials. Anesthesia affects the ERG and VEP in different degrees with different types and depths of anesthesia. Lighter level of anesthesia has less effect. One must coordinate with the anesthesiologist to attain a light level of anesthesia. When recording VEPs, the patient's EEG can show whether the patient is lightly anesthetized. If EEG amplitude is low, then ask the anesthesiologist to lighten anesthetic depth and wait a few minutes.

Anesthesiologists may start with fairly deep anesthesia. While anesthesiologist transitions from mask to intravenous anesthesia, I ask the anesthesiologist: "When the patient is stable will you please lighten the anesthesia as much as you are comfortable?" Ask permission before touching the patient. This act is especially important if you wish to raise the head to place the occipital electrode. Raising the head may crimp the tracheal tube.

If recording both ERGs and VEPs under anesthesia, then record the ERG first because it is more resistant to anesthesia. In the minutes it takes for the ERG recording, the patient's anesthetic depth lightens further. Rarely must one wait more minutes for the patient to lighten anesthetic depth. Oral glucose or sucrose produces analgesic effects for procedures in infants up to 18 months of age including recording ERGs [8].

Multifocal VEP

An important development in VEPs is the multifocal VEP (mfVEP) (Fig. 2.6). Erich Sutter adapted the mathematical sequences called binary m-sequences, creating a program that can extract hundreds of VEPs from occipital scalp. Yu et al. developed its first clinical application [9]. This program allows assessment of VEP activity in over 100 "channels" of each optic pathway. The common stimulus for mfVEPs is a dartboard pattern in which each sector is a contrast-reversing checkerboard pattern (Fig. 2.7). The individual waveform of mfVEP is similar to that of pattern VEP consisting of N1, P1, and N2 peaks [10].

Conventional VEPs evaluate the optic nerves and central pathways as a whole, in which small or localized optic pathway dysfunction is often not detectable. Using mfVEP, smaller dysfunctional areas may be detected with up to hundreds of stimulations presented in the same amount of time used to record a single whole nerve VEP with conventional methods [9, 11–13]. mfVEP can better evaluate asymmetry of visual function caused by optic nerve dysfunction or by localized retinal damage. Pathology can sometimes be detected that may be missed with a conventional VEP. However, it is limited to the central 60° of the visual field.

A common mfVEP 4-electrode montage is a negative electrode 2 cm below the inion and a positive active electrode 3–4 cm above the inion on the midline, and laterally place two positive active electrodes 4 cm off the midline at the level of inion (Fig. 2.8). All three positive active electrodes are connected to the positive

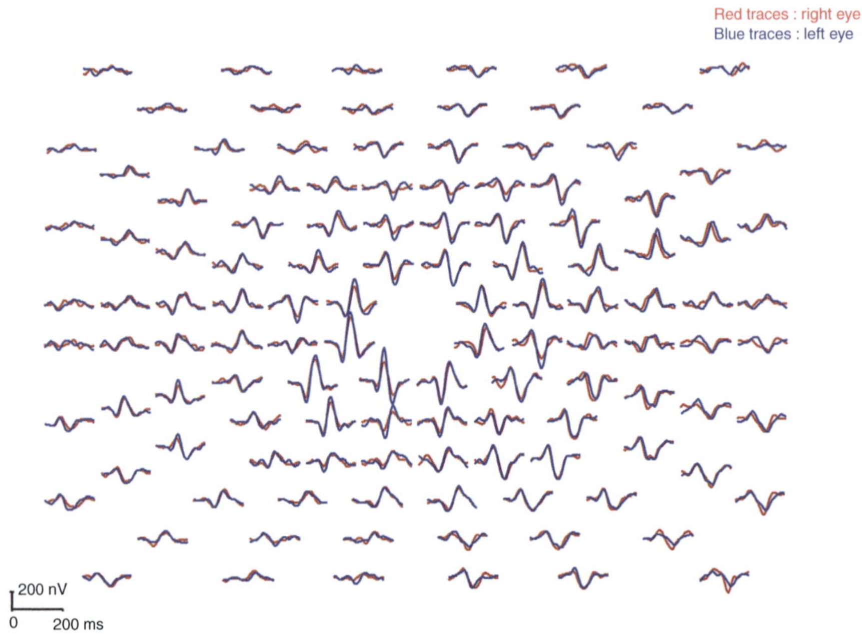

Fig. 2.6 Typical normal mfVEP traces. Red traces are from right eye. Blue traces are from left eye

Fig. 2.7 Dartboard pattern of stimulation for mfVEP test

input of the amplifier. The inion is the most prominent projection of the occipital bone at the posteroinferior (lower rear) part of the skull. The electrode below the inion usually serves as reference to the other three active electrodes. In other laboratories, the mfVEP is recorded with only two electrodes along the scalp midline, an electrode is at inion serving as the reference negative location, and a positive electrode is placed 3–4 cm above the inion with the ground electrode on the forehead.

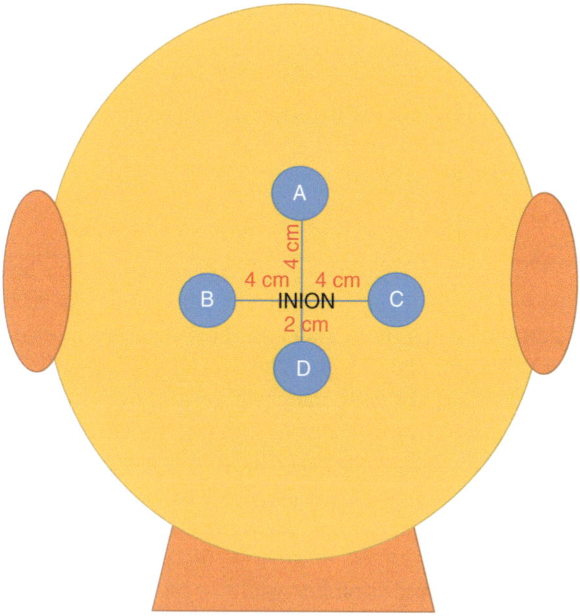

Fig. 2.8 Electrode placement for mfVEP test

Clinical Protocol

VEPs are useful to quantitate nerve dysfunction from retina to visual cortex. VEPs are recorded using a number of methods to evaluate the function of visual pathway. The VEP stimuli and protocol are chosen based on the patient's symptoms, history, and other information available. Abnormalities seen in VEPs are associated with the symptoms, but different diseases may have similar abnormal VEP. In another words, VEP abnormalities do not specify etiology. For example, retinal disease alone will drastically alter VEP parameters that are similar in an optic nerve disease. Interpretation of VEPs must be considered within the context of the patient's clinical appearance and information available from other tests and examinations. If the history and ophthalmic examination of a patient raises the possibility of dysfunctional retina, optic nerves, tracts, or cortices, then the VEP might yield diagnostically useful information. Patients are commonly referred with the question of whether their visual problems are due to retinal or central visual dysfunction. Recording both electroretinogram and VEP will usually answer this question.

When recording VEPs, the diagnostic yield is dependent upon appropriate selection of visual stimuli. If pattern stimuli are to be used, then it is important that a patient is tested with refractive error corrected. VEPs should be recorded with undilated pupils so that visual acuity is maximized. Refractive errors will affect the interpretation of the VEP results. The patients with LASIK or intraocular lens replacement may have one eye for distance vision and another for near vision. When VEPs are recorded in these patients, attention should be paid to the correction of the refractive error. Select the most powerful stimulus that allows the clearest pattern stimulation and the best cooperation. If the corrected visual acuity of the patient is approximately 20/200 (6/60) or better, and the patient does not have nystagmus, then the stimulus of first choice would be pattern reversal starting with about a 1-degree check size. If the resulting VEP appears well formed, then next use a smaller check size in the range of about 15′–20′ of arc. When estimating visual acuity, progressively smaller check sizes (e.g., 15′) and lower contrast (<50%) are recommended.

Check sizes subtending 10′–20′ of arc are the best stimuli for testing foveal vision. Checks of 40′–60′ of arc are superior in evaluating parafoveal function. Therefore, using more than one check size increases diagnostic yield. If the resulting VEP using the 1-degree check is poorly formed, then retest using a larger check size (2° of arc) or use a pattern onset stimulus. Pattern-reversal stimuli are the best choice in cooperative patients with better visual acuity, particularly when testing for possible effects of optic neuritis, hydrocephalus, ventriculitis, cortical hematomas, optic atrophy, neurofibromatosis, or compression of the optic pathways. However, poor macular function can cause poor VEPs. If the corrected acuity of the patient is approximately 20/200 to 20/400 (6/60–6/120) and/or the patient has nystagmus, then the preferred stimulus would be pattern onset.

References

1. Youssofzadeh V, et al. Signal propagation in the human visual pathways: an effective connectivity analysis. J Neurosci. 2015;35(39):13501–10.
2. Towle VL, et al. Locating VEP equivalent dipoles in magnetic resonance images. Int J Neurosci. 1995;80(1–4):105–16.
3. Slotnick SD, et al. Using multi-stimulus VEP source localization to obtain a retinotopic map of human primary visual cortex. Clin Neurophysiol. 1999;110(10):1793–800.
4. Di Russo F, et al. Cortical sources of the early components of the visual evoked potential. Hum Brain Mapp. 2002;15(2):95–111.
5. Jasper HH. Report of committee on methods of clinical examination in electroencephalography. Electroencephalogr Clin Neurophysiol. 1958;10:370–5.
6. Odom JV, et al. ISCEV standard for clinical visual evoked potentials: (2016 update). Doc Ophthalmol. 2016;133(1):1–9.
7. Hoffmann MB, Straube S, Bach M. Pattern-onset stimulation boosts central multifocal VEP responses. J Vis. 2003;3(6):432–9.
8. Pasek TA, Huber JM. Hospitalized infants who hurt: a sweet solution with oral sucrose. Crit Care Nurse. 2012;32(1):61–9.
9. Yu M, Brown B, Edwards MH. Investigation of multifocal visual evoked potential in anisometropic and esotropic amblyopes. Invest Ophthalmol Vis Sci. 1998;39(11):2033–40.
10. Sutter EE. Noninvasive testing methods: multifocal electrophysiology. In: Dartt DA, editor. Encyclopedia of the eye. Oxford: Academic Press; 2010.
11. Klistorner AI, et al. Multifocal topographic visual evoked potential: improving objective detection of local visual field defects. Invest Ophthalmol Vis Sci. 1998;39(6):937–50.
12. Yu MZ, Brown B. Variation of topographic visually evoked potentials across the visual field. Ophthalmic Physiol Opt. 1997;17(1):25–31.
13. Hood DC, et al. Tracking the recovery of local optic nerve function after optic neuritis: a multifocal VEP study. Invest Ophthalmol Vis Sci. 2000;41(12):4032–8.

Chapter 3
Electrooculography

Donnell Creel and Minzhong Yu

Origin of EOG

Elwin Marg described and named the electrooculogram in 1951 [1], and Geoffrey Arden [2] developed the first clinical application. With the cornea constantly positive, movement of the eye produces a shift of this electrical potential. The electrooculogram measures the potential that exists between the cornea and Bruch's membrane of retinal pigment epithelium (RPE) at the back of the eye. The potential produces a dipole field with the cornea a few millivolts positive compared to the back of the eye in an illuminated room. Although the origin of the EOG is the RPE, the light rise of the potential is associated with the RPE and rod photoreceptors. During dark adaptation, this potential reduces and reaches a minimum in about 10 minutes. This point in EOG is called dark trough (DT). After the light onset, the potential increases and reaches a maximum (i.e., light peak or LP) in about 10 minutes and then reduces in amplitude with damped oscillations for up to 90 minutes during the light adaption (Fig. 3.1).

The LP is associated with depolarization of the basal RPE membrane via one or more Cl channels [3–6]. Previous studies suggested that a calcium-sensitive chloride channel is the generator of LP, because the LP reduced in the mutations of some subunits of L-type voltage-gated calcium channels in mice [7, 8] and in the wild-type mice administered with the inhibitors of L-type voltage-gated calcium channels. The intracellular-free calcium is released from the endoplasmic reticulum (ER), which is regulated by L-type calcium channels in the basolateral membrane and ER bestrophin. Intracellular calcium then opens the basolateral

D. Creel (✉)
Moran Eye Center, University of Utah School of Medicine, Salt Lake City, UT, USA
e-mail: donnell.creel@hsc.utah.edu

M. Yu
Department of Ophthalmology, University Hospitals Eye Institute, Cleveland, OH, USA

M. Yu et al. (eds.), *Handbook of Clinical Electrophysiology of Vision*,
https://doi.org/10.1007/978-3-030-30417-1_3

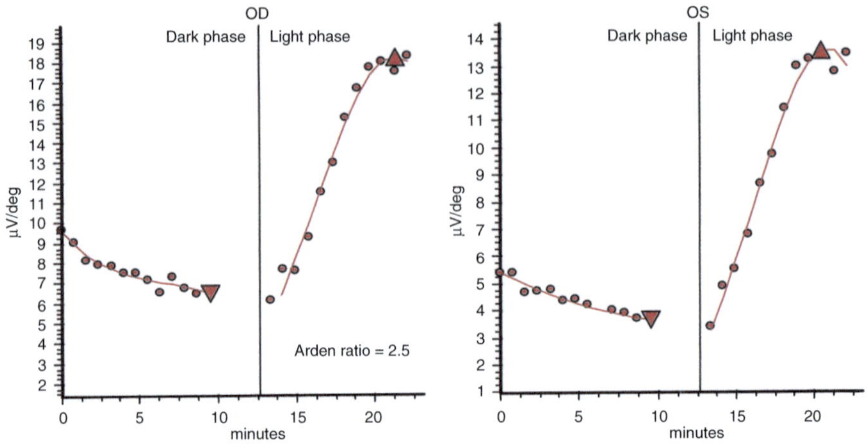

Fig. 3.1 Typical dark–light EOG measurements in normal subject. The unit of the EOG amplitude is μV/degree. Because the angle of the stimulation is 30 degrees, the absolute amplitude in μV can be derived by multiplying 30 degrees

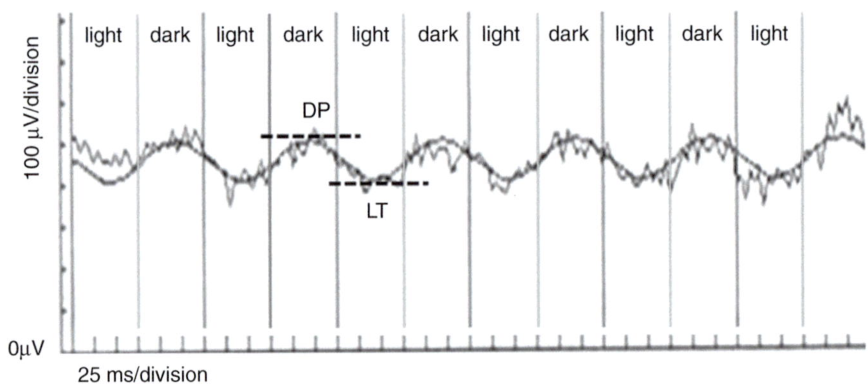

Fig. 3.2 A typical waveform of fast oscillation (Adapted from Fig. 5 of ISCEV Standard for clinical electrooculography (2017 update))

calcium-dependent chloride channel, increases the chloride conductance, depolarizes the basolateral membrane, and increases the transepithelial potential of RPE, which is recorded as LP.

Fast oscillation (FO) of EOG can also be generated with fast on- and off-cycles of the stimulating light (Fig. 3.2). When the light is on for 1 minute and off for another minute, fast oscillation (FO) is generated. Opposed to the conventional EOG, the standing potential reduces when the light is on, which causes light trough. The standing potential increases when the light is off, that elicits dark peak. The mechanism of the FO was studied by Steinberg laboratory in the 1980s. When the light is on, light-evoked reduction of subretinal potassium ion induces outward potassium current across RPE apical membrane and also reduces the transportation of chloride ions into the RPE that causes basolateral membrane to hyperpolarize and decreases the transepithelial potential. Then the FO trough is generated. When the light is off in a short interval, the transepithelial potential gradually increases to the maximum (i.e., dark peak) in about 35–45 seconds due to the ionic homeostasis is restored. The alternation between light and dark generates stable FO. Its amplitude can reflect the ionic permeability and electrical coupling of the RPE basal and apical membranes and can be used for the examination of RPE function [9].

Placement of Recording Electrodes, Parameters of Amplifier, and Stimulation

Skin electrodes are attached near the inner and outer canthus of each eye, and a ground electrode is usually attached to the forehead or an earlobe (Fig. 3.3).

The amplifiers should have a band pass from 0.1 to 30 Hz to make sure the waveforms of the saccadic movements are square waves. If the low-cut of the band pass is lower than 0.1 Hz, then the baseline drift may make the recording difficult. However, if the low-cut frequency is too high (e.g., 0.5 Hz), the tops of the square waves are not flat enough which makes the measurement not accurate.

The potential is measured by having the subject move his or her eyes horizontally a set angle. Ganzfeld for stimulation includes a chin rest to reduce head movement. The visual targets are two small red fixation lights 30 degrees apart that alternate on and off inside a Ganzfeld or on a screen in front of the patient (Fig. 3.4). With the saccadic angle of 30°, the typical peak-to-peak amplitude is between 250 and 1000 μV. The sampling frequency of the recording should be 1 kHz or higher in each channel for obtaining signal components up to 500 Hz. See Constable et al. (2017) for detailed ISCEV International Standard EOG protocol [10].

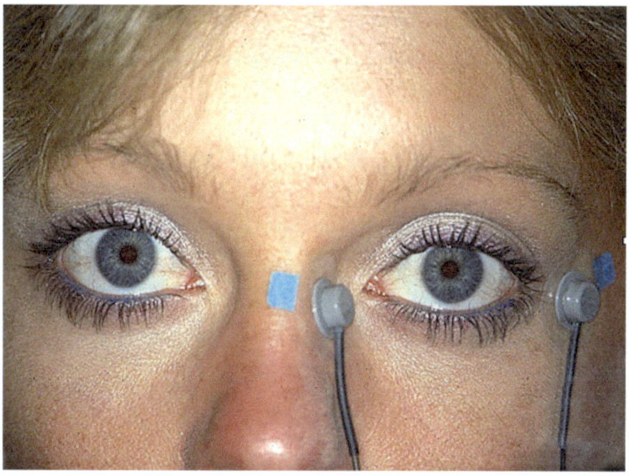

Fig. 3.3 EOG electrode placement

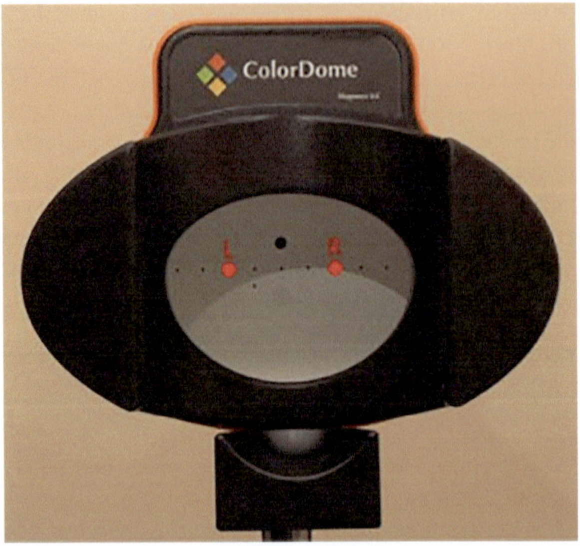

Fig. 3.4 Ganzfeld with LED lights exaggerated

Recording Procedure

The patient should be light adapted at least 10 minutes before the EOG test with the pupils dilated. After the electrodes are attached near inner and outer canthi, the procedure is explained to the patient, and the patient is asked to practice several times while baseline data are recorded. The procedure is simple. The patient keeps their head still while moving the eyes horizontally between the two red lights. The movement of the eyes produces a voltage swing between the electrodes on each side of the eye that is stored in the memory of a computer.

For recording LP:DT ratio, the background light is turned off after light adaptation and the patient training. A sample of EOG is taken about once a minute. The patient is asked to look horizontally between the two lights for five cycles in 10 seconds. The voltages induced by the saccades become smaller in the dark and reach its lowest potential, the so-called "dark trough", after 8–12 minutes. After 10–15 minutes in the dark phase, the background light with 100 photopic $cd{\cdot}m^{-2}$ is turned on, and the patient is again asked to do the same saccadic eye movement every minute as in the dark phase. Then, the voltages induced by the saccades rise and reach the maximum (i.e., light peak) in 7–12 minutes followed by the reduction in amplitude. The averaged voltages induced by the saccades in the five cycles are obtained at light peak and dark trough, respectively. And the light peak:dark trough ratio (LP:DT ratio), also called "Arden ratio", is obtained. The normal LP:DT ratio is about 2:1 or greater (Fig. 3.1). For individuals less than 60 years old, the ratio is considered abnormal if it is less than 1.8 and for over 60 years if ratio less than 1.7.

For recording FO, the similar method of the conventional EOG recording is used except light adaptation is not required, because FO is not affected by preadaptation [9–11]. The only difference is that the light–dark cycle is much shorter (e.g., about 2 minutes) in FO recording. The FO waveform is similar to a sinusoidal waveform (Fig. 3.2) [12]. The peak-to-trough amplitude, the ratio as well as the phase can be used as the clinical indexes [12]. The number of cycles used should be at least four.

Besides monitoring the function of RPE, EOG electrode placement is also used to monitor eye movements while the background light is kept constant [13–15].

References

1. Marg E. Development of electro-oculography; standing potential of the eye in registration of eye movement. AMA Arch Ophthalmol. 1951;45(2):169–85.
2. Arden GB, Barrada A, Kelsey JH. New clinical test of retinal function based upon the standing potential of the eye. Br J Ophthalmol. 1962;46(8):449–67.
3. Steinberg RH. Interactions between the retinal pigment epithelium and the neural retina. Doc Ophthalmol. 1985;60(4):327–46.
4. Gallemore RP, Steinberg RH. Effects of DIDS on the chick retinal pigment epithelium. I. Membrane potentials, apparent resistances, and mechanisms. J Neurosci. 1989;9(6):1968–76.
5. Gallemore RP, Steinberg RH. Light-evoked modulation of basolateral membrane Cl- conductance in chick retinal pigment epithelium: the light peak and fast oscillation. J Neurophysiol. 1993;70(4):1669–80.
6. Linsenmeier RA, Steinberg RH. Origin and sensitivity of the light peak in the intact cat eye. J Physiol. 1982;331:653–73.
7. Marmorstein LY, et al. The light peak of the electroretinogram is dependent on voltage-gated calcium channels and antagonized by bestrophin (best-1). J Gen Physiol. 2006;127(5):577–89.
8. Wu J, et al. Voltage-dependent calcium channel CaV1.3 subunits regulate the light peak of the electroretinogram. J Neurophysiol. 2007;97(5):3731–5.
9. Constable PA, et al. ISCEV Standard for clinical electro-oculography (2017 update). Doc Ophthalmol. 2017;134(1):1–9.
10. Constable PA, et al. Erratum to: ISCEV Standard for clinical electro-oculography (2017 update). Doc Ophthalmol. 2017;134(2):155.
11. Constable PA, et al. Correction to: ISCEV Standard for clinical electro-oculography (2017 update). Doc Ophthalmol. 2018;136(2):155.
12. Schneck ME, Shupenko L, Adams AJ. The fast oscillation of the EOG in diabetes with and without mild retinopathy. Doc Ophthalmol. 2008;116(3):231–6.
13. Yu MZ, Wu LZ, Wu DZ. Model parameters of the smooth pursuit eye movement system with electrooculogram. Doc Ophthalmol. 1990;76(1):37–46.
14. Danchaivijitr C, Kennard C. Diplopia and eye movement disorders. J Neurol Neurosurg Psychiatry. 2004;75 Suppl 4:iv24–31.
15. Rivaud-Pechoux S, et al. Longitudinal ocular motor study in corticobasal degeneration and progressive supranuclear palsy. Neurology. 2000;54(5):1029–32.

Part II
Clinical Application of Electrophysiology of Vision

Chapter 4
Congenital Non-Degenerative Retinal Diseases

Wajiha Jurdi Kheir, Roberto Gattegna, Minzhong Yu, Alessandro Racioppi, Alfonso Senatore, Donnell Creel, and Alessandro Iannaccone

This entity, congenital non-degenerative retinal diseases, is comprised of a heterogeneous group of disorders characterized by the following:

1. Congenital impairment of visual function
2. Lack of (or minimal) progression of visual function impairment
3. Abnormal development of the retina and/or other ocular structures

These conditions can be due to a defect in the development or function of rods or rod micro-circuitry (inner retinal neurons), cones or cone micro-circuitry (inner retinal neurons), or retinal pigment epithelium (RPE). Patients mainly complain of night blindness, photophobia, reduced visual acuity, and nystagmus. Ophthalmoscopy findings can vary from an absolutely normal looking fundus examination to abnormalities in the macula and in pigmentation. Patients may exhibit foveal hypoplasia associated with abnormal ocular coherence tomography (OCT) findings as well as

W. J. Kheir · A. Senatore · A. Iannaccone
Center for Retinal Degenerations and Ophthalmic Genetic Diseases, Duke University School of Medicine, Duke Eye Center, Department of Ophthalmology, Durham, NC, USA

R. Gattegna
Center for Retinal Degenerations and Ophthalmic Genetic Diseases, Duke University School of Medicine, Duke Eye Center, Department of Ophthalmology, Durham, NC, USA

Retina Service, Israelitic Hospital, Rome, Italy

M. Yu (✉)
Department of Ophthalmology, University Hospitals Eye Institute, Cleveland, OH, USA
e-mail: minzhong.yu@uhhospitals.org

A. Racioppi
Center for Retinal Degenerations and Ophthalmic Genetic Diseases, Duke University School of Medicine, Duke Eye Center, Department of Ophthalmology, Durham, NC, USA

University of North Carolina, Chapel Hill, NC, USA

D. Creel
Moran Eye Center, University of Utah School of Medicine, Salt Lake City, UT, USA

M. Yu et al. (eds.), *Handbook of Clinical Electrophysiology of Vision*,
https://doi.org/10.1007/978-3-030-30417-1_4

abnormal retinal pigmentation including but not limited to hypopigmentation, retinal pigment deposits, and anomalous retinal reflexes.

Night-Blinding Disorders (Fig. 4.1)

Congenital Stationary Night Blindness

Congenital stationary night blindness (CSNB) is a type of retinal disease with impaired night vision and may exhibit X-linked, recessive, or dominant inheritance patterns. It can also be classified into two groups based on ERG results. In the Schubert–Bornschein type, the b-wave amplitude is lower than the a-wave amplitude (electronegative mixed, rod–cone dark-adapted ERG), whereas in the Riggs type, both the a- and b-waves of the mixed ERG are reduced. The Schubert–Bornschein type of CSNB usually presents with congenital nystagmus, decreased visual acuity, and myopia, whereas the patients with Riggs-type CSNB usually have normal visual acuity and do not exhibit nystagmus. The dominant form of CSNB is also known as the Nougaret type, and is similar to the Riggs type, and due to mutations in the *GNAT1* gene that encodes for the rod transducin alpha subunit [1].

The Schubert–Bornschein CSNB can be further subdivided into "complete" and "incomplete" forms [2]. In the "complete" form of CSNB (type 1 CSNB or CSNB1), there is little to no measurable rod-driven ERG. CSNB1 is associated with mutations in the following genes: *NYX* [3, 4], *GRM6* [5, 6], *TRPM1* [7–9],

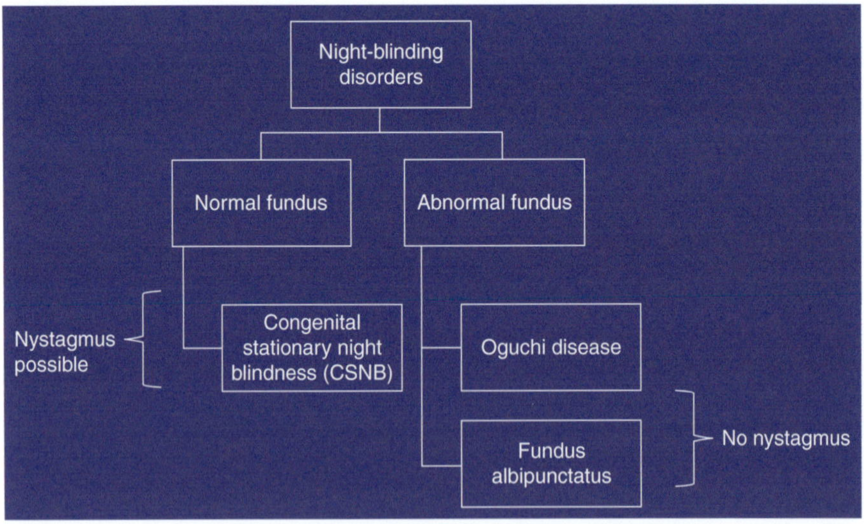

Fig. 4.1 Classification of congenital nonprogressive night blinding disorders. All these patients experience night blindness and can present either with normal or abnormal fundus examinations and presence or absence of associated nystagmus

GPR179 [10, 11], *GNB3* [12], *LRIT3* [13], and *SLC24A1* [2]. In the "incomplete" form of CSNB (type 2 CSNB or CSNB2), the electronegative mixed ERG response waveform remains present, but both the rod-driven and the cone-driven responses are decreased but measurable [14]. CSNB2 is associated with mutations in the *CACNA1F* [15, 16] and *CABP4* genes [17, 18].

The differences of ffERG between CSNB1 and CSNB2 are summarized as follows (Fig. 4.2): In CSNB1, the rod response of ERG is diminished, the mixed rod–cone response of ERG shows significant decrease of b-wave amplitude while the a-wave amplitude is not decreased significantly, and the photopic cone response is slightly decreased or delayed. In CSNB2, the b-wave amplitude of rod response of ERG is reduced but recordable, the mixed rod–cone response of ERG shows mildly reduced b-wave amplitude while the a-wave amplitude is also mildly reduced, and the photopic cone response is significantly decreased [14, 19–21].

Oguchi Disease

Oguchi disease, first reported by Oguchi in Japan in 1907, is an autosomal recessive form of congenital stationary night blindness characterized by "Mizuo–Nakamura phenomenon," a golden or gray-white discoloration of the retina that disappears and returns to normal fundus appearance after several hours of dark adaptation and shows up again quickly after the eye receives light (Fig. 4.3). This disease is classified into two types depending on the genetic mutation: type 1 is characterized by mutations in the genes coding for arrestin (*SAG* or *ARR*) and type 2 by mutations in genes coding for rhodopsin kinase (*RHOK*) [22–28]. Rhodopsin kinase and arrestin are important in the visual transduction. After the light activates rhodopsin and hyperpolarizes rod cells, rhodopsin kinase is released from recoverin and is activated. The released rhodopsin kinase phosphorylates rhodopsin. Arrestin can only bind to phosphorylated rhodopsin and prevents the rhodopsin from binding to transducin, terminating phototransduction. In Oguchi disease, abnormal rhodopsin kinase or arrestin significantly slows recovery kinetics of rod and cone phototransduction [29]. The patients with Oguchi disease have impaired dark adaptation, but the rod sensitivity to light can be improved after a lengthy dark adaptation.

The typical features of the ffERG of Oguchi disease include markedly diminished to almost undetectable standard dark-adapted b-wave amplitude, marked a-wave amplitude reduction and even more profound b-wave reduction to the mixed rod–cone dark-adapted ffERG, and either normal or variably decreased standard light-adapted ffERG b-wave amplitude and flicker ERG amplitude [30–33] (Fig. 4.3). It has been reported that nonstandard testing methods, such as the repetitive-flash ffERG protocol in which double- or triple-flash stimulations after prolonged dark adaptation are used to elicit ERG response, can help identify more atypical cases of Oguchi disease [34]. In addition, the mfERG may show attenuated and prolonged P1 peak response densities in the peripheral portion of the stimulated area, while usually the foveal peak responses remain normal [35].

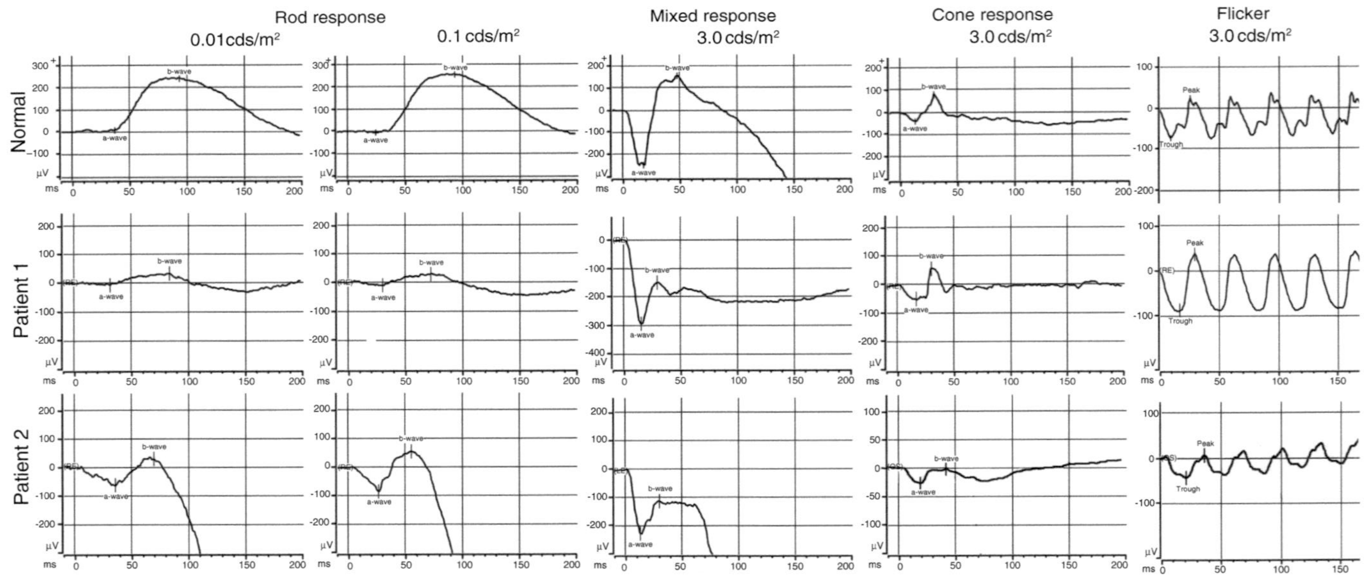

Fig. 4.2 Electroretinographic findings in CSNB. Rod, mixed, cone transient, and flicker responses are presented for a 10-year-old Caucasian boy (Patient 1) with a mutation in the *TRPM1* gene causing CSNB1 and another 10-year-old Caucasian boy (Patient 2) with a mutation in *CACNA1F* causing CSNB2. Both children had congenital nystagmus and reduced acuity with a complaint of poor visual performance in dimly lit environments. The molecular diagnoses corroborated the ffERG findings that showed electronegative mixed ffERGs in both cases, but associated with severely reduced rod responses in Patient 1 with CSNB1 and partial preservation thereof in Patient 2 who shows also reduction of the cone-driven ffERGs, as it is typical for CSNB2 instead

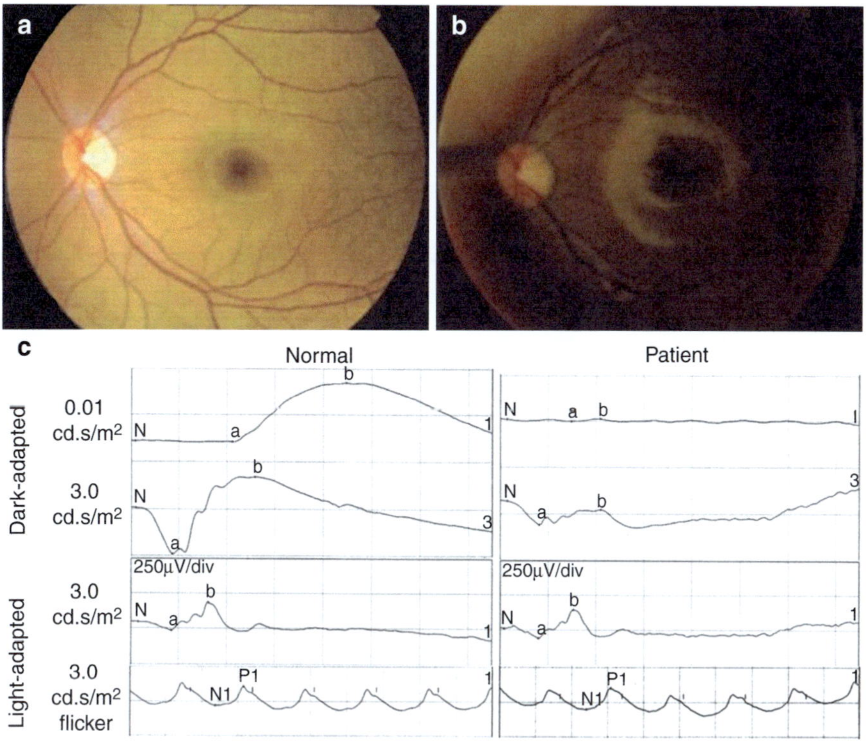

Fig. 4.3 Clinical and functional findings in Oguchi disease. Fundus photo showing "Mizuo–Nakamura phenomenon" before (**a**) and after (**b**) dark adaptation. (**c**) Full-field electroretinograms (ERGs) of a normal person and a patient with Oguchi disease. The rod b-waves were nearly undetectable. The mixed ffERGs exhibited an electronegative morphology, and the cone ffERGs were near normal (Adapted from Figure 1 and 2 [30])

Fundus Albipunctatus

Fundus albipunctatus (FA) is a congenital night-blinding disorder characterized by the presence of small, white retinal dots usually seen mainly in the mid-peripheral area but can extend past the arcades into the macular region corresponding with hyperreflective, discrete deposits at the level of the RPE, displacing the overlying photoreceptor outer segments (Fig. 4.4). SD-OCT can show reduced length of the photoreceptor outer segments in patient with fundus albipunctatus in areas overlying the deposits [36, 37]. Fundus albipunctatus is mainly associated with mutations in the *RDH5* gene that encodes the 11-*cis* retinol dehydrogenase, a key enzyme in the visual cycle [38].

By definition, the rod ffERG in fundus albipunctatus is reduced, and it has been shown that greater compromise of the cone-mediated ffERG tends to be associated with greater reductions in the rod-mediated ffERGs as well [35]. In the dark-adapted mixed rod–cone response, the a- and b-wave amplitudes are decreased after standard dark adaptation periods, but can be improved up to normal range after a

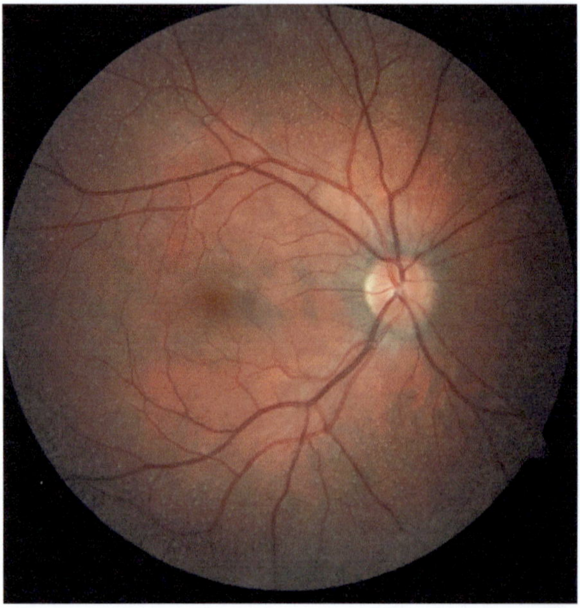

Fig. 4.4 Clinical findings in fundus albipunctatus. This is a fundus photograph of a 23-year-old Caucasian male patient with a complaint of lifelong vision difficulties in dimly lit environments and prolonged dark adaptation, who was found to have compound heterozygous *RDH5* mutations, diagnostic for fundus albipunctatus. Upon fundus examination, there are discrete punctate white dots disseminated throughout the mid-periphery with sparing of the macular region. The remainder of the retinal examination is normal

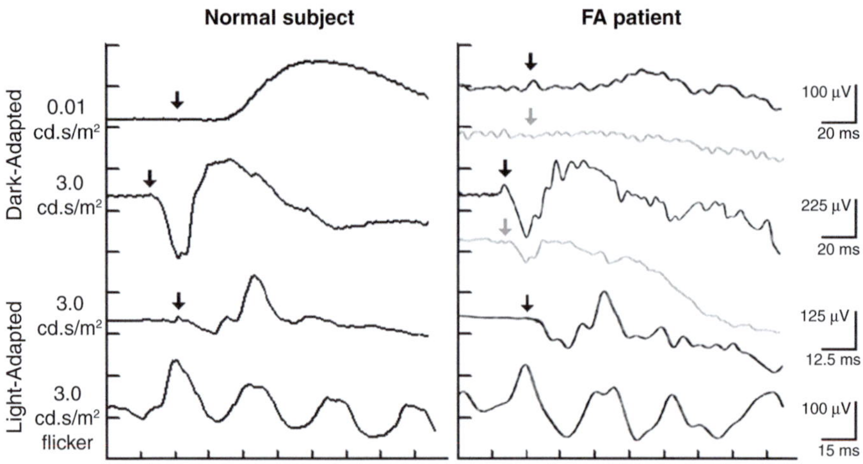

Fig. 4.5 Electroretinographic findings in fundus albipunctatus. The ffERG of a 6-year-old Hispanic girl with compound heterozygous *RDH5* mutations is presented, illustrating the change in dark-adapted response dynamics with prolonged dark adaptation. After standard (30 min) dark adaptation, the rod response (gray traces) is nonrecordable, and the mixed response is markedly reduced and electronegative. After 120-min dark adaptation (dark traces), the rod response has greatly improved, and the mixed response is also much larger and no longer exhibits the previously noted electronegative morphology. The light-adapted cone responses were normal. These findings are typical for patients with fundus albipunctatus

prolonged dark adaptation of around 120 min (Fig. 4.5) [36, 37, 39–51]. The light-adapted ERG b-wave amplitude that reflects the function of cone pathway in fundus albipunctatus varies from normal range to significantly reduced. Older fundus albipunctatus patients tend to have more decrease of light-adapted ERG b-wave than younger ones [52]. In patients that exhibit cone function abnormalities, the mfERG can show amplitude reductions in corresponding areas that also show loss of sensitivity on visual field testing and pathological changes on OCT imaging [36, 42].

Photophobia Disorders

Photophobia disorders (Fig. 4.6) are characterized by nystagmus, impaired color discrimination, and reduced visual acuity. Although some degree of foveal hypoplasia is usually a feature in all cases, the fundus may appear to be normal or hypopigmented.

Achromatopsia (ACHM) and Blue Cone Monochromatism (BCM)

Achromatopsia (rod monochromatism) is an autosomal recessive photophobia disorder presenting during infancy with poor visual acuity (usually 20/200 or less), congenital nystagmus that tends to abate over time, and minimal to completely

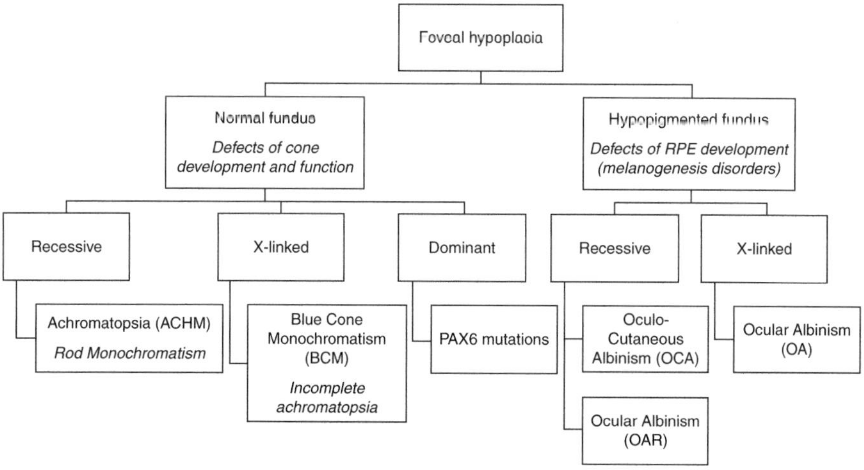

Fig. 4.6 Classification of congenital nonprogressive photophobia disorders. In these conditions, all patients have some degree of clinically detectable foveal hypoplasia and can have either normal fundus examinations or pigmentation deficits

absent color vision. The prevalence is approximately 1 in 30,000 people, and it is caused by mutations in genes responsible for the cone conduction cascade: *CNGA3*, *CNGB3*, *GNAT2*, *PDE6C*, *PDE6H*, and *ATF6*. Mutations in CNGB3 account for 40–50% of the cases of ACHM [53, 54].

Blue cone monochromatism (BCM) is a form of incomplete achromatopsia inherited as an X-linked trait that is characterized by photophobia, a red-green color vision defect, with sparing of blue cone-mediated function, congenital nystagmus that tends to abate over time, and decreased visual acuity (can at times reduce to 20/80). BCM is less common than ACHM, with an estimated prevalence of about 1:100,000, although this figure may represent an underestimate due to frequent misdiagnosis and complexities of the molecular genetic testing needed to confirm the diagnosis. It is mainly caused by deletions in genes expressing red/L (long wavelength) and green/M (middle wavelength) cone photoreceptor opsins: *OPN1LW/OPN1MW* gene cluster, or in the *LCR* region that controls the expression of these two genes [55]. BCM variants are also caused by point mutations (e.g., the C203R mutation) and by complex haplotypes including multiple variants at various positions (e.g., LVAVA, LIAVA, MVAVA). These rarer variants tend to cause phenotypic manifestations that are more similar to those of an X-linked cone dystrophy, and female carriers can also exhibit a phenotype not usually seen in the typical BCM families [56–59].

There are some important differences between ACHM and BCM that are otherwise hardly distinguishable from one another clinically. While both cause color vision defects, there is complete or near-complete lack of color vision in ACHM while the blue color axis by definition is preserved in BCM patients. This can be reflected by their respective ffERG responses. Both ACHM patients and BCM patients will usually have normal rod responses and normal to near-normal mixed dark-adapted ffERG responses, usually with a reduced a-wave due to lack of cone contribution to the response, accompanied by severely diminished or absent light-adapted cone-driven responses to either transient or flicker stimuli. In BCM, there is sometimes residual light-adapted cone-mediated responses (Fig. 4.7). However, it is also not uncommon to see molecularly confirmed ACHM patients present with residual light-adapted cone-mediated responses as well. The refractive defect also usually differs between ACHM and BCM: There is usually a hyperopic refractive error in ACHM and a moderate to high myopic refractive error in BCM. Because of the mode of inheritance, ACHM patients may have affected siblings but otherwise a negative family history, and both males and females can be affected. Unlike them, BCM affects only males, and these patients will usually have positive family history among siblings or maternal relatives. However, cases without any additional family history definitely exist. Another finding shared by ACHM and BCM is the presence of foveal cavitation lesions in OCT (Fig. 4.8). In general, while extremely useful to pose the diagnosis of either ACHM or BCM, no ERG test can conclusively and reliably distinguish between the two conditions, and psychophysical color vision testing is ultimately more informative. Chromatic visual field testing has also been

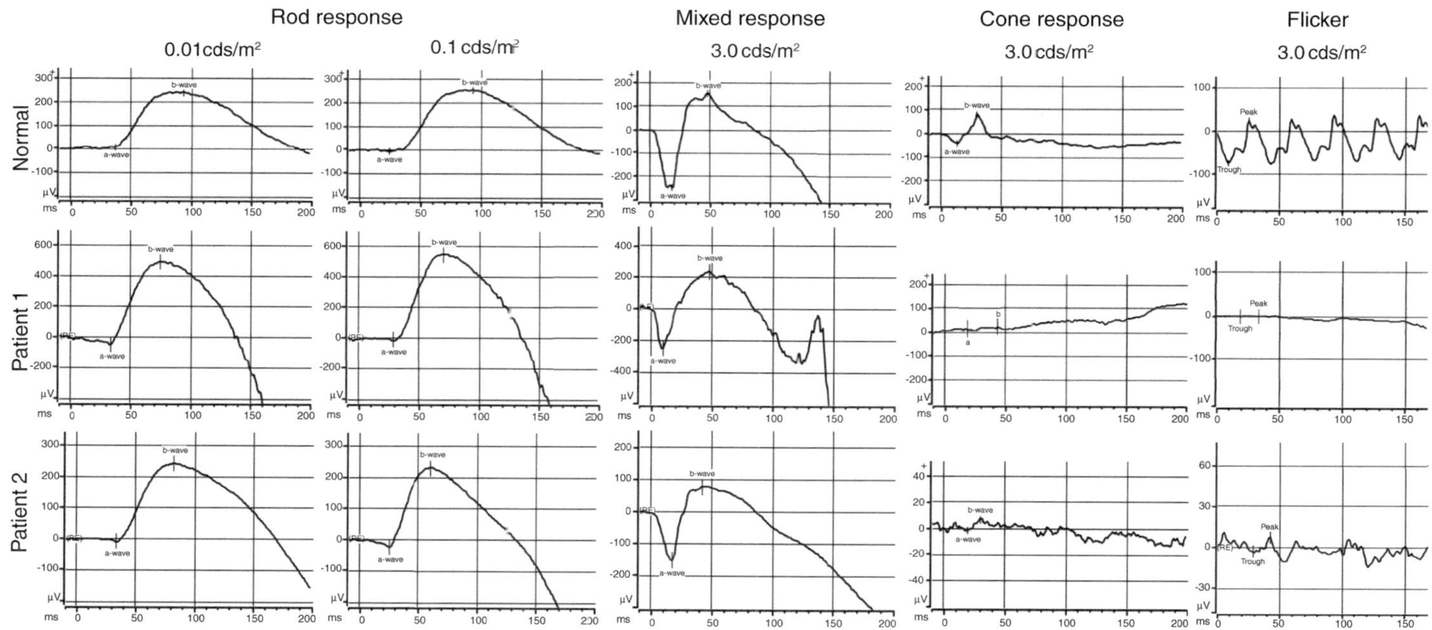

Fig. 4.7 Electroretinographic findings in achromatopsia and blue cone monochromatism. Standard rod, mixed, transient cone, and flicker responses are shown for a case of achromatopsia, an 8-year-old Caucasian girl compound heterozygous for a nonsense mutation and a complex deletion with a frameshift effect, both leading to truncation of the *CNGB3* gene product (Patient 1) and a case with blue cone monochromatism (BCM), a 60-year-old Caucasian male hemizygous for a *LCR* deletion (Patient 2). Both patients presented with lifelong history of photophobia, reduced acuity, poor color vision, and nystagmus at birth. Their dark-adapted rod and mixed ERGs exhibit normal responses (except for mildly reduced mixed a-waves, in the achromat more than in the blue cone monochromat), which is attributable to the lack of cone contribution to the response. In the patient with achromatopsia, all light-adapted responses are essentially nonrecordable, whereas in the patient with BCM, there is some residual light-adapted response detected despite his advanced age

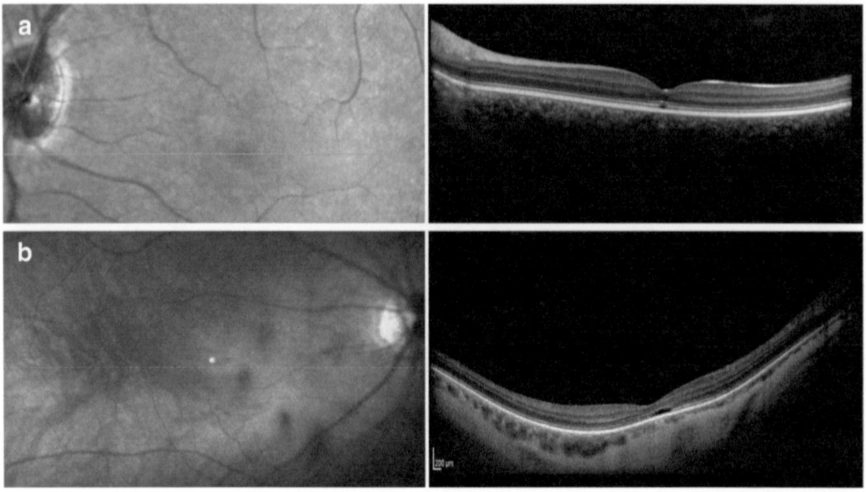

Fig. 4.8 Retinal imaging findings in achromatopsia and blue cone monochromatism. Macular SD-OCT scans from the same patients shown in Fig. 4.7 show foveal cavitation lesions resulting from focal foveal ellipsoid zone (EZ) gaps in a patient with (**a**) achromatopsia and (**b**) blue cone monochromatism (BCM). Note the staphylomatous ectasia of the macula of the BCM patient, consistent with his high myopic refractive status, a common finding in BCM not typically present in achromatopsia

shown to reliably distinguish between the two conditions [60], and chromatic stimuli are relied upon as outcome measures for future treatment trials [55].

Phenotypes Associated with Mutations in the Paired Box 6 Gene

The paired box gene 6 *(PAX6)* is a master regulator of the central nervous system and ocular development [61]. There are more than 330 reported mutations in the *PAX6* gene, leading to a varied pattern and severity of autosomal dominant inherited phenotypes [62]. These mutations are not only best known for causing aniridia but can also cause anterior segment dysgenesis, keratopathy, glaucoma, cataract, ectopia lentis, foveal hypoplasia, and nystagmus [63, 64]. Photophobia in these patients is usually associated with measurable reductions in cone-driven ERG responses [64–67]. Atypical cases of congenital nystagmus and photophobia without classical/overt aniridia and subtle anterior segment changes presenting with dominant inheritance and reduced cone ERGs can be due to PAX6 mutations [65] (Fig. 4.9). This finding attests to the fact that the photophobia, that is the hallmark of the aniridia, is not merely due to the lack of iris pigmentation, but is the expression of a developmental retinal defect primarily affecting the cone-driven pathways.

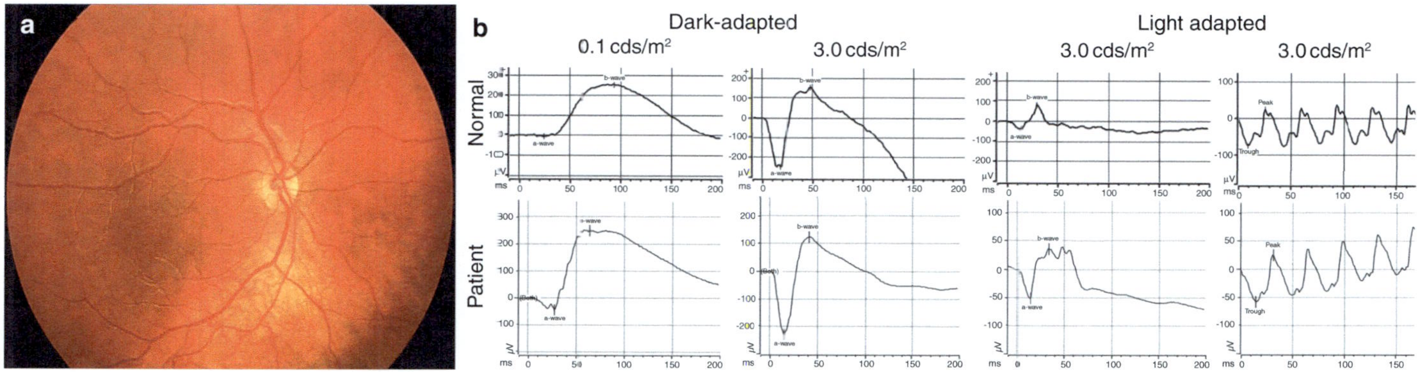

Fig. 4.9 Clinical and electroretinographic findings in PAX6-related disease. (**a**) Fundus examination in a 25-year-old Caucasian male patient with severe photophobia, congenital nystagmus, and reduced acuity associated with a frameshift *PAX6* mutation showed an essentially normal examination except for absent foveal reflexes due to foveal hypoplasia. His anterior segment examination also revealed posterior embryotoxon, corneal pannus, and ectropion uveae. (**b**) His ffERG showed mild attenuation of light-adapted responses compared to a normal control. Essentially the same findings were documented in his daughter, who harbored the same *PAX6* mutation (not shown)

References

1. Dryja TP, et al. Missense mutation in the gene encoding the alpha subunit of rod transducin in the Nougaret form of congenital stationary night blindness. Nat Genet. 1996;13(3):358–60.
2. Riazuddin SA, et al. A mutation in SLC24A1 implicated in autosomal-recessive congenital stationary night blindness. Am J Hum Genet. 2010;87(4):523–31.
3. Pusch CM, et al. The complete form of X-linked congenital stationary night blindness is caused by mutations in a gene encoding a leucine-rich repeat protein. Nat Genet. 2000;26(3):324–7.
4. Bech-Hansen NT, et al. Mutations in NYX, encoding the leucine-rich proteoglycan nyctalopin, cause X-linked complete congenital stationary night blindness. Nat Genet. 2000;26(3):319–23.
5. Zeitz C, et al. Mutations in GRM6 cause autosomal recessive congenital stationary night blindness with a distinctive scotopic 15-Hz flicker electroretinogram. Invest Ophthalmol Vis Sci. 2005;46(11):4328–35.
6. Dryja TP, et al. Night blindness and abnormal cone electroretinogram ON responses in patients with mutations in the GRM6 gene encoding mGluR6. Proc Natl Acad Sci U S A. 2005;102(13):4884–9.
7. Audo I, et al. TRPM1 is mutated in patients with autosomal-recessive complete congenital stationary night blindness. Am J Hum Genet. 2009;85(5):720–9.
8. Li Z, et al. Recessive mutations of the gene TRPM1 abrogate ON bipolar cell function and cause complete congenital stationary night blindness in humans. Am J Hum Genet. 2009;85(5):711–9.
9. van Genderen MM, et al. Mutations in TRPM1 are a common cause of complete congenital stationary night blindness. Am J Hum Genet. 2009;85(5):730–6.
10. Audo I, et al. Whole-exome sequencing identifies mutations in GPR179 leading to autosomal-recessive complete congenital stationary night blindness. Am J Hum Genet. 2012;90(2):321–30.
11. Peachey NS, et al. GPR179 is required for depolarizing bipolar cell function and is mutated in autosomal-recessive complete congenital stationary night blindness. Am J Hum Genet. 2012;90(2):331–9.
12. Vincent A, et al. Biallelic mutations in GNB3 cause a unique form of autosomal-recessive congenital stationary night blindness. Am J Hum Genet. 2016;98(5):1011–9.
13. Zeitz C, et al. Whole-exome sequencing identifies LRIT3 mutations as a cause of autosomal-recessive complete congenital stationary night blindness. Am J Hum Genet. 2013;92(1):67–75.
14. Glass IA, et al. Genetic mapping of a cone and rod dysfunction (Aland Island eye disease) to the proximal short arm of the human X chromosome. J Med Genet. 1993;30(12):1044–50.
15. Bech-Hansen NT, et al. Loss-of-function mutations in a calcium-channel alpha1-subunit gene in Xp11.23 cause incomplete X-linked congenital stationary night blindness. Nat Genet. 1998;19(3):264–7.
16. Strom TM, et al. An L-type calcium-channel gene mutated in incomplete X-linked congenital stationary night blindness. Nat Genet. 1998;19(3):260–3.
17. Zeitz C, et al. Mutations in CABP4, the gene encoding the Ca2+-binding protein 4, cause autosomal recessive night blindness. Am J Hum Genet. 2006;79(4):657–67.
18. Littink KW, et al. A novel homozygous nonsense mutation in CABP4 causes congenital cone-rod synaptic disorder. Invest Ophthalmol Vis Sci. 2009;50(5):2344–50.
19. Bijveld MM, et al. Genotype and phenotype of 101 dutch patients with congenital stationary night blindness. Ophthalmology. 2013;120(10):2072–81.
20. Kurata K, Hosono K, Hotta Y. Long-term clinical course in a patient with complete congenital stationary night blindness. Case Rep Ophthalmol. 2017;8(1):237–44.
21. Miyake Y, et al. Congenital stationary night blindness with negative electroretinogram. A new classification. Arch Ophthalmol. 1986;104(7):1013–20.
22. Hayashi T, et al. A novel homozygous GRK1 mutation (P391H) in 2 siblings with Oguchi disease with markedly reduced cone responses. Ophthalmology. 2007;114(1):134–41.
23. Yamamoto S, et al. Defects in the rhodopsin kinase gene in the Oguchi form of stationary night blindness. Nat Genet. 1997;15(2):175–8.

24. Mucciolo DP, et al. A novel GRK1 mutation in an Italian patient with Oguchi disease. Ophthalmic Genet. 2018;39(1):137–8.
25. Fuchs S, et al. A homozygous 1-base pair deletion in the arrestin gene is a frequent cause of Oguchi disease in Japanese. Nat Genet. 1995;10(3):360–2.
26. Maw M, et al. Two Indian siblings with Oguchi disease are homozygous for an arrestin mutation encoding premature termination. Hum Mutat. 1998;11(Suppl 1):S317–9.
27. Maw MA, et al. Oguchi disease: suggestion of linkage to markers on chromosome 2q. J Med Genet. 1995;32(5):396–8.
28. Nakazawa M, et al. Oguchi disease: phenotypic characteristics of patients with the frequent 1147delA mutation in the arrestin gene. Retina. 1997;17(1):17–22.
29. Cideciyan AV, et al. Null mutation in the rhodopsin kinase gene slows recovery kinetics of rod and cone phototransduction in man. Proc Natl Acad Sci U S A. 1998;95(1):328–33.
30. Huang L, et al. A Chinese family with Oguchi's disease due to compound heterozygosity including a novel deletion in the arrestin gene. Mol Vis. 2012;18:528–36.
31. Sergouniotis PI, et al. Mizuo-Nakamura phenomenon in Oguchi disease due to a homozygous nonsense mutation in the SAG gene. Eye (Lond). 2011;25(8):1098–101.
32. Yuan A, Nusinowitz S, Sarraf D. Mizuo--Nakamura phenomenon with a negative waveform ERG. Br J Ophthalmol. 2011;95(1):147–8. 156
33. Miyake Y, et al. Electrophysiological findings in patients with Oguchi's disease. Jpn J Ophthalmol. 1996;40(4):511–9.
34. Fujinami K, et al. Oguchi disease with unusual findings associated with a heterozygous mutation in the SAG gene. Arch Ophthalmol. 2011;129(10):1375–6.
35. Hayashi T, et al. Macular dysfunction in Oguchi disease with the frequent mutation 1147delA in the SAG gene. Ophthalmic Res. 2011;46(4):175–80.
36. Schatz P, et al. Fundus albipunctatus associated with compound heterozygous mutations in RPE65. Ophthalmology. 2011;118(5):888–94.
37. Liu X, et al. RDH5 retinopathy (fundus albipunctatus) with preserved rod function. Retina. 2015;35(3):582–9.
38. Yamamoto H, et al. Mutations in the gene encoding 11-cis retinol dehydrogenase cause delayed dark adaptation and fundus albipunctatus. Nat Genet. 1999;22(2):188–91.
39. Margolis S, Siegel IM, Ripps H. Variable expressivity in fundus albipunctatus. Ophthalmology. 1987;94(11):1416–22.
40. Iannaccone A, et al. Fundus albipunctatus in a 6-year old girl due to compound heterozygous mutations in the RDH5 gene. Doc Ophthalmol. 2007;115(2):111–6.
41. Hajali M, et al. Diagnosis in a patient with fundus albipunctatus and atypical fundus changes. Doc Ophthalmol. 2009;118(3):233–8.
42. Ruther K, et al. Clinical and genetic findings in a patient with fundus albipunctatus. Ophthalmologe. 2004;101(2):177–85.
43. Hotta K, et al. Macular dystrophy in a Japanese family with fundus albipunctatus. Am J Ophthalmol. 2003;135(6):917–9.
44. Nakamura M, Miyake Y. Macular dystrophy in a 9-year-old boy with fundus albipunctatus. Am J Ophthalmol. 2002;133(2):278–80.
45. Miyazaki K, et al. [A case of fundus albipunctatus with a retinol dehydrogenase 5 gene mutation in a child]. Nippon Ganka Gakkai Zasshi. 2001;105(8):530–4.
46. Naz S, et al. Mutations in RLBP1 associated with fundus albipunctatus in consanguineous Pakistani families. Br J Ophthalmol. 2011;95(7):1019–24.
47. Dryja TP. Molecular genetics of Oguchi disease, fundus albipunctatus, and other forms of stationary night blindness: LVII Edward Jackson Memorial Lecture. Am J Ophthalmol. 2000;130(5):547–63.
48. Skorczyk-Werner A, et al. Fundus albipunctatus: review of the literature and report of a novel RDH5 gene mutation affecting the invariant tyrosine (p.Tyr175Phe). J Appl Genet. 2015;56(3):317–27.
49. Wang NK, et al. Multimodal fundus imaging in fundus albipunctatus with RDH5 mutation: a newly identified compound heterozygous mutation and review of the literature. Doc Ophthalmol. 2012;125(1):51–62.

50. Pras E, et al. Fundus albipunctatus: novel mutations and phenotypic description of Israeli patients. Mol Vis. 2012;18:1712–8.
51. Sergouniotis PI, et al. Phenotypic variability in RDH5 retinopathy (Fundus Albipunctatus). Ophthalmology. 2011;118(8):1661–70.
52. Niwa Y, et al. Cone and rod dysfunction in fundus albipunctatus with RDH5 mutation: an electrophysiological study. Invest Ophthalmol Vis Sci. 2005;46(4):1480–5.
53. Pascual-Camps I, et al. Diagnosis and treatment options for achromatopsia: a review of the literature. J Pediatr Ophthalmol Strabismus. 2018;55(2):85–92.
54. Zobor D, Zobor G, Kohl S. Achromatopsia: on the doorstep of a possible therapy. Ophthalmic Res. 2015;54(2):103–8.
55. Luo X, et al. Blue cone monochromacy: visual function and efficacy outcome measures for clinical trials. PLoS One. 2015;10(4):e0125700.
56. Gardner JC, et al. Blue cone monochromacy: causative mutations and associated phenotypes. Mol Vis. 2009;15:876–84.
57. Sumaroka A, et al. Blue cone monochromacy caused by the C203R missense mutation or large deletion mutations. Invest Ophthalmol Vis Sci. 2018;59(15):5762–72.
58. Gardner JC, et al. Three different cone opsin gene array mutational mechanisms with genotype-phenotype correlation and functional investigation of cone opsin variants. Hum Mutat. 2014;35(11):1354–62.
59. Orosz O, et al. Myopia and late-onset progressive cone dystrophy associate to LVAVA/MVAVA exon 3 interchange haplotypes of opsin genes on chromosome X. Invest Ophthalmol Vis Sci. 2017;58(3):1834–42.
60. Skalak C, et al. A simple, clinician-friendly perimetric approach to the differential diagnosis between blue cone monochromacy (BCM) and achromatopsia (ACHM): a pilot study. Invest Ophthalmol Vis Sci. 2018;59(9):4045.
61. Walther C, Gruss P. Pax-6, a murine paired box gene, is expressed in the developing CNS. Development. 1991;113(4):1435–49.
62. Hingorani M, Hanson I, van Heyningen V. Aniridia. Eur J Hum Genet. 2012;20(10):1011–7.
63. Guo H, et al. A large novel deletion downstream of PAX6 gene in a Chinese family with ocular coloboma. PLoS One. 2013;8(12):e83073.
64. Jia X, et al. A novel mutation of PAX6 in Chinese patients with new clinical features of Peters' anomaly. Mol Vis. 2010;16:676–81.
65. Hood MP, et al. Abnormal cone ERGs in a family with congenital nystagmus and photophobia harboring a p.X423Lfs mutation in the PAX6 gene. Doc Ophthalmol. 2015;130(2):157–64.
66. Yokoi T, et al. Genotype-phenotype correlation of PAX6 gene mutations in aniridia. Hum Genome Var. 2016;3:15052.
67. Tremblay F, et al. Effects of PAX6 mutations on retinal function: an electroretinographic study. Am J Ophthalmol. 1998;126(2):211–8.

Chapter 5
Macular Dystrophies

Wajiha Jurdi Kheir, Minzhong Yu, Alfonso Senatore, Roberto Gattegna, Alessandro Racioppi, Donnell Creel, and Alessandro Iannaccone

Stargardt's Disease/Fundus Flavimaculatus

Stargardt's disease (STGD) was first described by Dr. Stargardt in the early 1900s and is the most common inherited form of macular dystrophy, estimated to affect 1:8,000–1:10,000 people [1]. Multiple genes are responsible for this disease. In its classical manifestation, STGD is an autosomal recessive disease (STGD1) caused by mutations in the *ABCA4* gene, previously known also as *ABCR* [2]. Several other genes can cause STGD-like macular dystrophies, usually inherited in autosomal dominant fashion, and these include the *ELOVL4* (STGD3) and the *PROM1* genes (STGD4) [3]. Mutations in the *PRPH2* gene (peripherin/RDS) can also cause phenotypes that might be clinically indistinguishable from STGD [4].

W. J. Kheir · A. Senatore · A. Iannaccone
Center for Retinal Degenerations and Ophthalmic Genetic Diseases, Duke University School of Medicine, Duke Eye Center, Department of Ophthalmology, Durham, NC, USA

M. Yu (✉)
Department of Ophthalmology, University Hospitals Eye Institute, Cleveland, OH, USA
e-mail: minzhong.yu@uhhospitals.org

R. Gattegna
Center for Retinal Degenerations and Ophthalmic Genetic Diseases, Duke University School of Medicine, Duke Eye Center, Department of Ophthalmology, Durham, NC, USA

Retina Service, Israelitic Hospital, Rome, Italy

A. Racioppi
Center for Retinal Degenerations and Ophthalmic Genetic Diseases, Duke University School of Medicine, Duke Eye Center, Department of Ophthalmology, Durham, NC, USA

University of North Carolina, Chapel Hill, NC, USA

D. Creel
Moran Eye Center, University of Utah School of Medicine, Salt Lake City, UT, USA

M. Yu et al. (eds.), *Handbook of Clinical Electrophysiology of Vision*,
https://doi.org/10.1007/978-3-030-30417-1_5

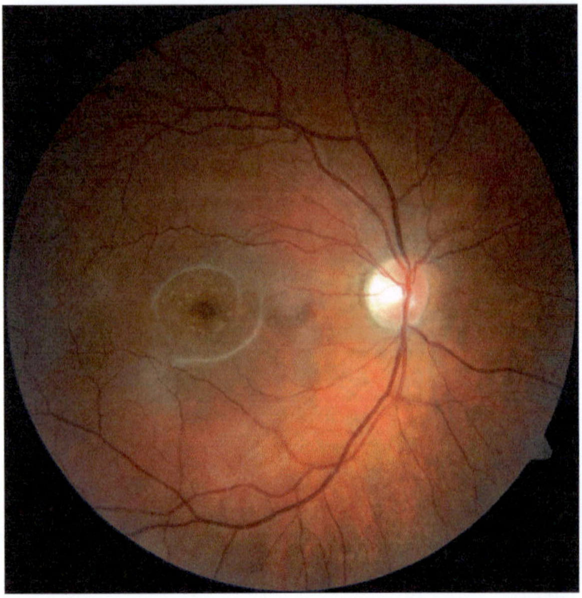

Fig. 5.1 Clinical description of Type I Stargardt's disease. Fundus photo of a patient with two confirmed *ABCA4* mutations with Type I Stargardt's disease, with disease manifestations limited to the foveal and perifoveal area. There are drusenoid flecks around the fovea with central thinning, loss of central foveal reflex, and RPE mottling. The remainder of the fundus examination is normal. This patient presented no other symptom except reduced visual acuity in her childhood

Most commonly, STGD becomes symptomatic in childhood and young adulthood, but numerous instances of late-onset disease, as late as in the sixth decade of life, have been reported. In its most typical manifestation, STGD shows gray-yellow flecks surrounding the fovea (Fig. 5.1), that can appear as pisciform or drusen-like deposits, surrounding an area of foveal thinning and beaten-bronze retinal reflexes. Flecks vary greatly in both size and abundance, and can be present as a small ring of deposits around the fovea or as disseminated, widespread deposits that might extend well beyond the vascular arcades. Taken together, these flecks configure an appearance known as fundus flavimaculatus that can rarely be also the only clinical phenotype with no macular lesions (Fig. 5.2). In virtually all cases, visual acuity becomes reduced as soon as there is significant foveal involvement varying widely in age of onset. Accessory symptoms can also include blurry central or pericentral vision and prolongation of dark-adaptation time, usually proportional to the extent of fleck accumulation due to a delay in the recycling of the visual pigments and more often present in advanced disease stages [5]. For *ABCA4*-related recessive disease, a clinical and electroretinographic classification scheme has been proposed [6], whereby Type I, delimited disease (as in Fig. 5.1) usually shows normal full-field flash (ff) ERG amplitudes but often exhibits delayed cone responses (qualitative anomaly), Type II (as shown in Fig. 5.2) disease usually exhibits mild to moderate cone or cone>rod amplitude losses, and Type III more severe atrophic changes and invariably with a more severe cone–rod dystrophic ffERG pattern. Examples of ffERGs meeting these classification criteria are shown in Fig. 5.3. A more refined disease sequence based on ophthalmoscopic, perimetric, imaging, and electroretinographic features has been characterized, showing lipofuscin accumulation in the RPE is an

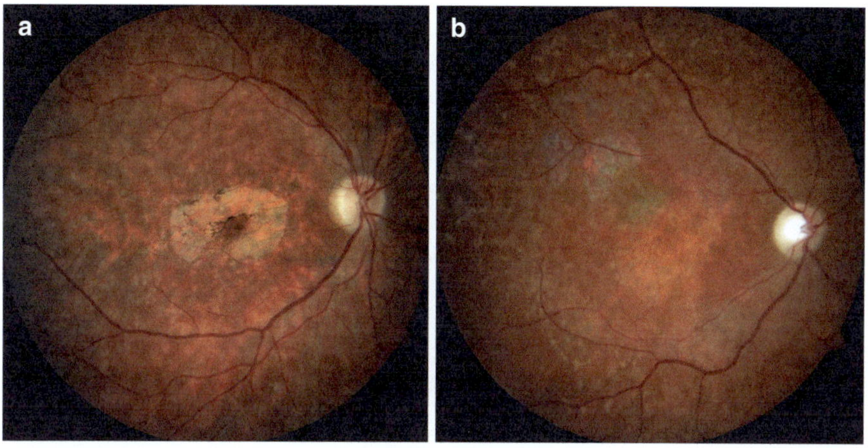

Fig. 5.2 Examples of more widespread forms of *ABCA4*-related Stargardt's disease with disseminated flecks (fundus flavimaculatus). (**a**) Fundus photo showing discrete central RPE atrophy and pigmentary changes and surrounding disseminated atypical grayish flecks. (**b**) Fundus photo showing patches of ill-defined RPE atrophy and retinal thinning, surrounded by a large halo of disseminated yellowish flecks extending beyond the arcades

essential, early-stage manifestation of the *ABCA4*-related disease process [5]. This phenomenon usually leads to progressive RPE loss and functional compromise of the overlying photoreceptors. The lipofuscin deposits often lead to a characteristic angiographic finding known as the dark choroid effect present in most but not all STGD patients (Fig. 5.4).

STGD patients can also be classified into four subtypes based on mfERG testing results [7]. As shown in Fig. 5.5, patients with type 1 disease have severely decreased mfERG response densities only in the foveal area; type 2 patients have diffuse decrease in mfERG response densities but residual detectable foveal peaks; type 3 patients have generalized depression of all responses densities across the macular region, usually associated with delayed responses as well; and type 4 patients have normal foveal mfERG response density in the foveal area but reduced and often delayed responses in the perifoveal regions. The latter category of patients usually exhibits later-onset phenotypes with prolonged foveal sparing.

Clinically, dominant STGD-like forms can be completely indistinguishable from recessive STGD. Despite their dominant inheritance, variable penetrance and expressivity can make other cases in the family hard to identify and far less symptomatic, whereby molecular genetic diagnostic testing can prove essential to recognize the dominant inheritance pattern. The clinical and imaging appearance as well as the severity of ffERG compromise can vary significantly between patients, even within the same family, usually as a function of disease severity. Examples of diseases associated with mutations in *ELOVL4* and *PROM1* are illustrated in Figs. 5.6 and 5.7, respectively. The ffERG abnormalities can range from milder functional phenotypes similar to Type I recessive STGD patients mainly characterized by

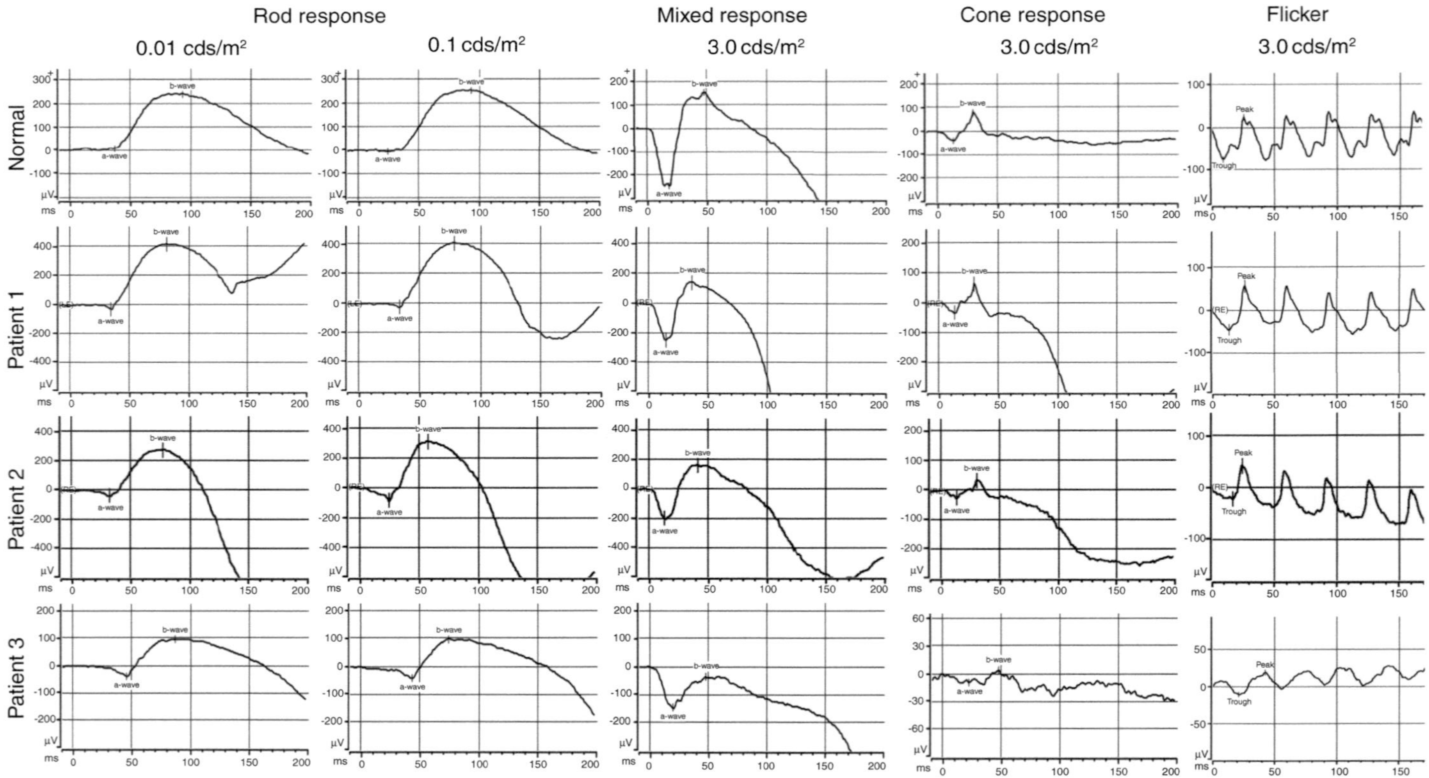

Fig. 5.3 Examples of ERGs from different types of ABCA4 Stargardt's disease. Patient 1 exhibits normal rod, mixed, and cone ERG responses consistent with Type I Stargardt's disease. Patient 2 exhibits normal rod and mixed responses and mildly diminished cone responses consistent with Type II Stargardt's disease. Patient 3 exhibits a cone–rod dystrophy (CORD) pattern, with cone ≫ rod diminished responses meeting the criteria for Type III Stargardt's disease. The latter patients represent a continuum of the *ABCA4*-related spectrum into the more widespread cone–rod dystrophies that are genetically heterogeneous

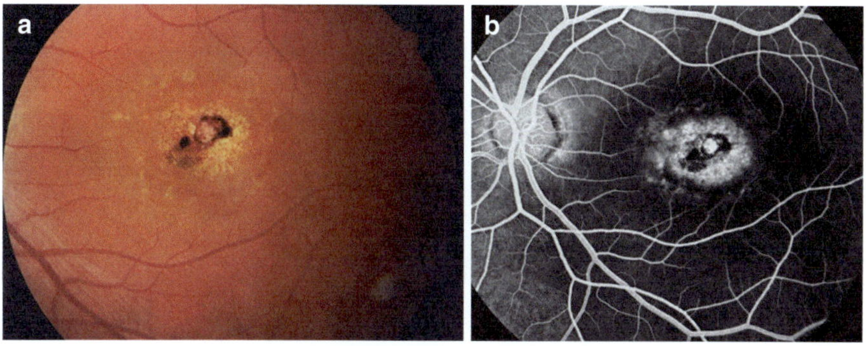

Fig. 5.4 Dark choroid effect in Stargardt's disease. (**a**) Fundus photos of a patient with *ABCA4*-positive Stargardt's disease showing central RPE atrophy, pigmentary changes, and perilesional yellowish flecks. (**b**) Fluorescein angiography shows staining of the flecks surrounded by a dark halo of hypofluorescence, configuring the so-called "dark choroid" effect often seen in Stargardt's disease and attributed to a blocking effect from lipofuscin deposits in the RPE

delayed cone-driven responses and a left-shift of the photopic hill (premature cone ERG b-wave amplitude declines with increasing flash intensities) to more severe, frank cone–rod dystrophy-type phenotypes. Thus, while very important for diagnostic and prognostic purposes, ffERG results do not allow a differentiation between recessive, *ABCA4*-related autosomal recessive Stargardt's disease (Fig. 5.3) and the dominant forms (Figs. 5.6 and 5.7).

X-Linked Retinoschisis

X-linked retinoschisis (XLRS) was first described in 1898 in two affected brothers by Dr. Josef Haas [8]. It is an inherited disease characterized by the splitting of the inner retina occurring almost exclusively in males in both eyes and usually starts in early childhood. XLRS affects 1 in 5,000–25,000 males in the population [9]. The vast majority of XLRS cases are caused by mutations or deletions in the *RS1* gene on the X chromosome. The *RS1* gene encodes retinoschisin, a protein secreted by photoreceptors essential for trans-synaptic signal transmission in the retina and maintenance of intraretinal cellular adhesion. Virtually all *RS1* mutations lead to defective secretion and intracellular retinoschisin retention [10]. Some cases of presumed XLRS patients who do not exhibit *RS1* mutations have been reported [11].

Visual acuity in XLRS is usually reduced early on due to macular retinoschisis and disruption of the integrity of the foveal photoreceptors. Acuity can become progressively reduced in childhood and adolescence with more gradual decline over time in adulthood [12]. If atrophy develops, then legal blindness will ensue [13]. XLRS can ultimately cause atrophic degeneration of the macular tissue, at which stage the typical cystic schitic macular changes may become no longer appreciable either clinically or by optical coherence tomography (OCT) criteria and can often disrupt the physical

a
b
c
d
e

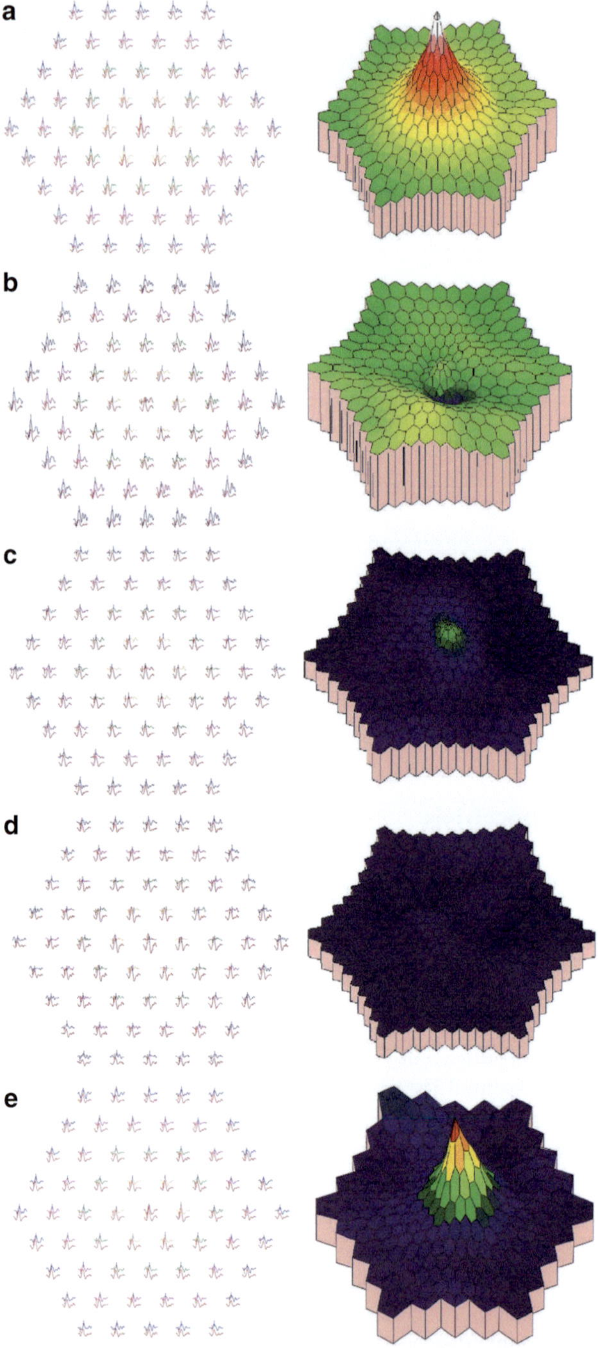

Fig. 5.5 Representative mfERG responses from various stages of Stargardt's disease . Trace arrays of mfERGs (left) and 3D plots of the response densities of the P1 component (right) are shown for a normal eye (**a**) and for eyes with type 1 (**b**), type 2 (**c**), type 3 (**d**), and type 4 (**e**) Stargardt's disease, all with confirmed *ABCA4* mutations (see text for further details about this mfERG-based proposed classification)

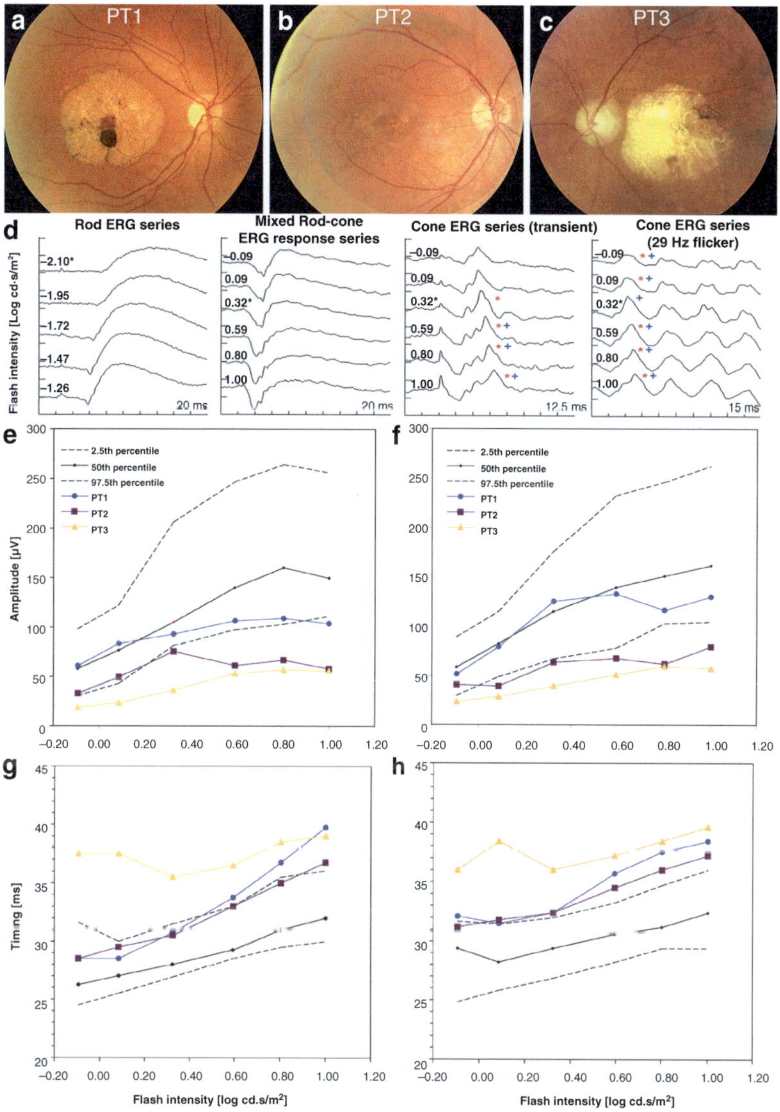

Fig. 5.6 Clinical and electroretinographic findings in *ELOVL4*-associated autosomal dominant Stargardt-like macular dystrophy. (**a–c**) Fundus photographs of two siblings in their late 40s and early 50s, respectively (PT1 and PT2), and their father in his early 80s (PT3) who had initially been diagnosed with age-related macular degeneration. They were later found to share same 5-base pair deletion (790-794delAACTT) in the *ELOVL4* gene. While PT1 and PT2 exhibit end-stage geographic atrophy-like changes, PT2 exhibits a milder form with retinal pigment epithelial changes and mild atrophy, reflected also in the less severe visual complaints. Note the fleck-like lesions at the edge of the macular atrophy of PT1 and similar but smaller lesions parafoveally also in PT2. (**d**) Dark- and light-adapted ffERG series of PT1, the least affected from a ffERG standpoint, showing mildly decreasing transient cone b-wave and flicker amplitudes and increasing delays with increasing flash intensities. (**e–h**) The intensity response plots of the ffERG cone transient b-waves (**e**, **g**) and flicker responses (**f**, **h**) are shown for both amplitudes and timing, showing various degrees of amplitude reduction and delays, more severe of all in PT3. Rod and mixed responses were within normal limits in PT1 and PT2, whereas mild rod response reductions were seen in PT3, configuring a late-stage CORD pattern (not shown)

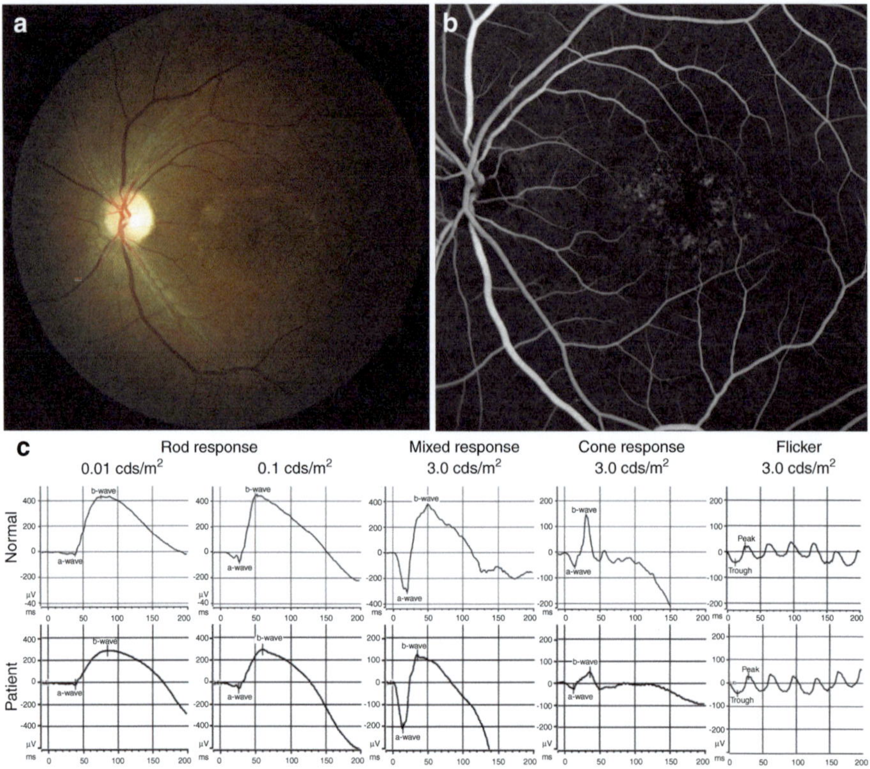

Fig. 5.7 Clinical and electroretinographic findings in Stargardt-like macular dystrophy linked to a *PROM1* mutation. (**a**) Fundus photo of a 63-year-old African-American male patient with complaint of dyschromatopsia, declining acuity, and difficulties with night driving, in whom a heterozygous 2-base pair *PROM1* deletion leading to a frameshift effect caused a Stargardt-like macular dystrophy. (**b**) His fluorescein angiogram shows staining of a ring of hyperfluorescent lesions around the fovea. (**c**) The ffERG shows normal rod and mixed responses, and moderate delay and mild cone amplitude loss. His sister shares the same *PROM1* deletion and is also affected, but with a more severe degree of both central and cone function involvement (not shown)

integrity of the peripheral outer retina as well (Fig. 5.8). At times, the schitic changes can be so widespread across the retina ("sponge cake-like" retinal appearance on OCT) that macular retinoschisis can be hard to appreciate clinically in enface views [14] but is well illustrated by OCT scans (Fig. 5.9). Severe complications, such as retinal detachment or vitreous hemorrhage, may occur in XLRS patients further reducing vision or, in the most severe of the cases, cause blindness. Retinal detachments can at times be present at or shortly after birth and include leukocoria, posing a significant differential diagnostic challenge with persistent primary hyperplastic vitreous, retinoblastoma, and other such conditions. Spontaneous reattachment of peripheral retinal detachments can also occur and can cause subretinal fibrotic changes and/or RP-like pigmentary changes, more commonly in streaks or limited to affected retinal sectors. Much milder and usually later-onset phenotypes are also possible, with no detectable schisis and

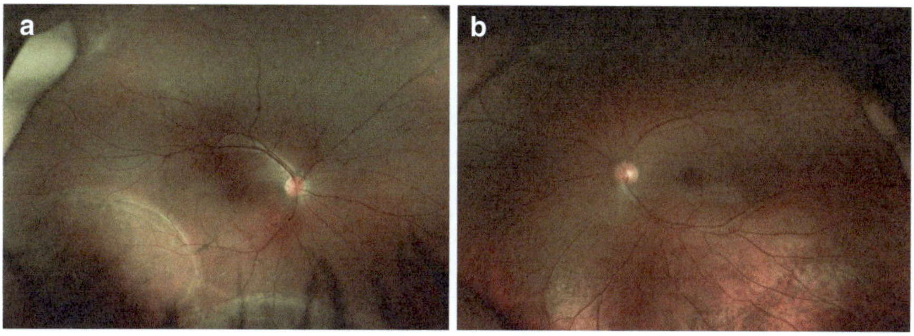

Fig. 5.8 Clinical findings in X-linked retinoschisis. In this 26-year-old Caucasian male patient with a confirmed *RS1* disease-causing hemizygous mutation and complaint of reduced visual acuity since early childhood, the fundus photograph of the right eye (**a**) shows two large areas of inferior and inferotemporal chronic peripheral bullous retinoschisis with fibrotic demarcation lines. (**b**) The left eye shows no peripheral retinoschisis but shows disseminated patchy areas of golden tapetal-like reflexes (TLR)

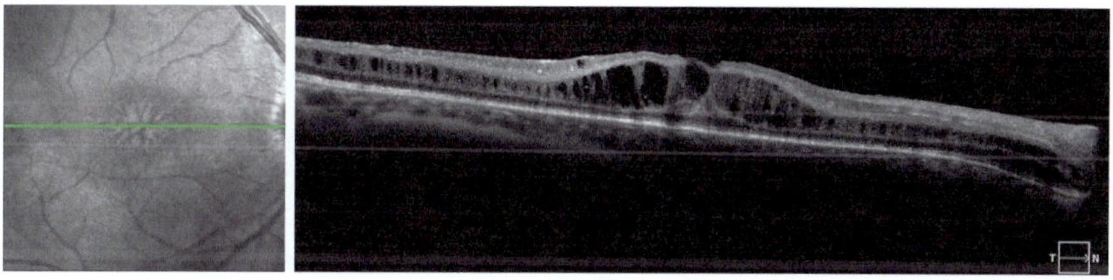

Fig. 5.9 Macular OCT findings in XLRS. SD-OCT scan of the same patient with XLRS shown in Fig. 5.8, showing disseminated hyporeflective macular intraretinal cystic spaces, especially in the foveal area, consistent with a typical case of macular schisis in XLRS

exhibiting only metallic, golden tapetal-like reflexes (TLRs) [15]. TLRs are also often observed in XLRS cases exhibiting macular and/or peripheral retinoschisis (Fig. 5.8).

While molecular genetic diagnostic testing for disease-causing changes in the *RS1* gene is ultimately necessary to confirm the suspicion that a patient is affected by XLRS, and macular OCT often exhibits cystic changes mimicking cystoid macular edema, ffERG testing is usually essential in diagnosing this condition, even in cases only with TLRs or other mild or unusual signs. The typical ffERG finding of XLRS is an electronegative mixed dark-adapted ERG response, due to a normal or near-normal a-wave followed by a variable degree of b-wave truncation (Fig. 5.10). Both pure rod-driven and cone-driven ERG responses can be variably reduced and delayed, and often the latter ones exhibit some degree of electronegativity as well. The flicker cone ERG often exhibits a split peak. All these abnormalities are usually associated with variable degrees of ON- and OFF-response compromise. It has also been reported b-wave electronegativity may be proportional to the extent of the TLRs when present [15]. A greater degree of ON-response compromise is usually associated with a greater degree of b-wave electronegativity. Progressive decline

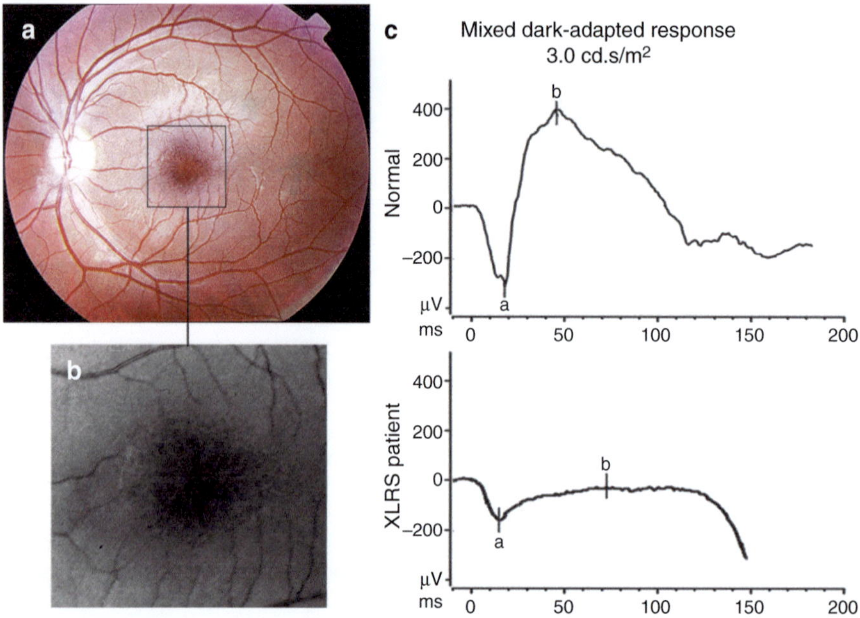

Fig. 5.10 Macular and electroretinographic characteristics of XLRS. (**a**) The macula of this 8-year-old Caucasian boy with a complaint of severe light aversion and acuity not correctable to better than 20/60 is remarkable for characteristic clustered polycystic foveal changes in a spoke–wheel pattern that, on red-free imaging. (**b**) confers a spongiform appearance of the macula. His mixed dark-adapted ffERG shows a typical electronegative waveform with a large a-wave but a markedly depressed b-wave. (**c**) His light-adapted ffERG responses were also markedly reduced (not shown). Molecular genetic testing revealed a R197C *RS1* mutation, also identified in his maternal uncle, who was also affected

of the a-wave has been reported, reflecting an increasing degree of photoreceptor functional compromise [9, 16, 17].

The mfERG is consistently abnormal in XLRS, exhibiting a wide range of possible abnormalities, ranging from mild to severe response density reductions [18, 19]. The mfERG changes appear to parallel to the severity of the macular retinoschisis. Examples of mfERGs from XLRS patients with confirmed RS1 mutations and their corresponding macular OCTs are illustrated in Fig. 5.11.

Vitelliform Macular Dystrophy (Best's Disease)

Vitelliform macular dystrophy, also called Best's disease, was first identified by Franz Best in 1905. It is an inherited disease due to a defect in the retinal pigment epithelium (RPE) affecting macular integrity. Best's disease is caused by mutations or deletions in the *BEST1* gene (also known as *VMD2*) expressed in the RPE [20, 21]. Most of the cases of vitelliform macular dystrophy are autosomal dominant in

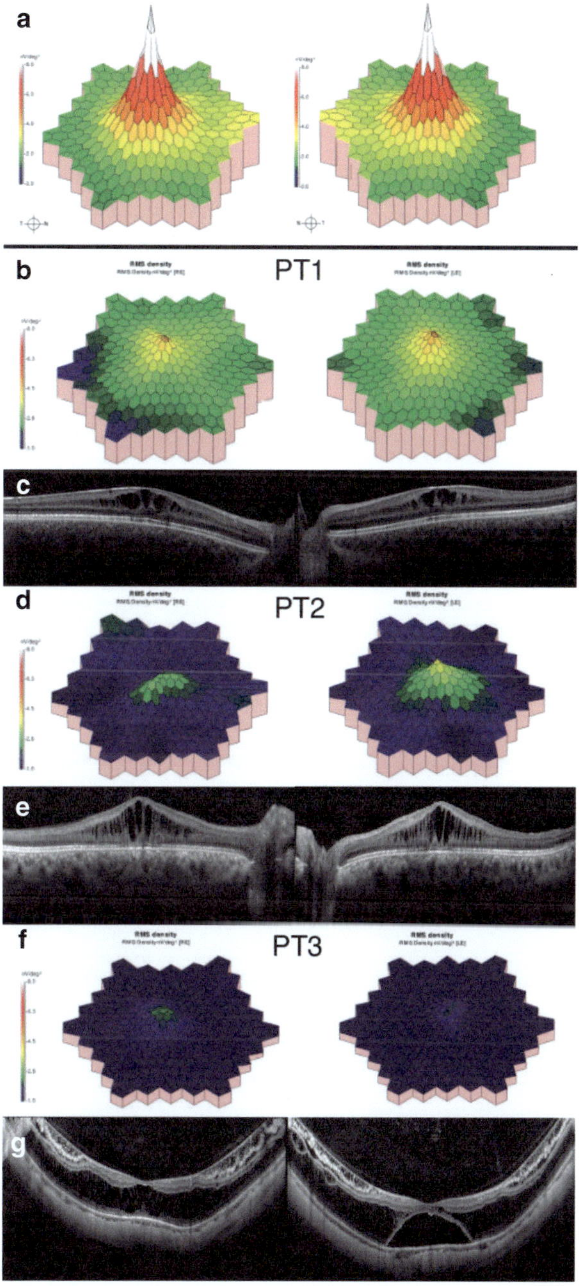

Fig. 5.11 Multifocal ERG macular and microanatomical findings in XLRS. Various levels of mfERG compromise are seen in XLRS and, to a certain degree, parallel to the extent of macular retinoschisis in affected patients. (**a**) Normal mfERG example as reference. (**b**, **c**) PT1: Mildly reduced mfERG foveal peak densities and SD-OCT scan in a case with mild macular retinoschisis confined to the foveal area. (**d**, **e**) PT2: Moderate generalized reduction in mfERG response densities and SD-OCT scan in a case with moderate macular retinoschisis. (**f**, **g**) PT3: Severely depressed mfERG responses and SD-OCT scan in a patient with severe inner and outer macular retinoschisis, complicated further by a bilateral myopic posterior staphyloma and a serous foveal detachment in the left eye

inheritance [22], while some cases are inherited in autosomal recessive fashion [23]. Disease-causing changes in the *BEST1* gene can also cause phenotypic expressions other than typical vitelliform ones, such as the spectrum of conditions known as the autosomal recessive bestrophinopathies (ARBs), autosomal dominant vitreoretinochoroidopathy (ADViRC), and retinitis pigmentosa [22, 24–26].

Best's disease can be clinically classified into several progressive stages:

- *Pre-vitelliform stage*: Usually no visible fundus abnormalities are observed, but RPE changes or a yellowish dot at fovea may be seen.
- *Vitelliform stage*: Most often observed in infancy and childhood. A sharply demarcated, round, subretinal yellowish lesion simulating an egg yolk-like appearance is classically appreciated, with a dense, localized hyperreflective subfoveal lesion seen on OCT (Figs. 5.12a, b and 5.13). At this stage, lesions can be extrafoveal and/or multiple. Satellite vitelliform lesions can be observed also outside of the vascular arcades. The size of the lesions and the amount of elevation on the retinal plane vary considerably, and spontaneous regression is possible. Many patients at this stage are still asymptomatic.

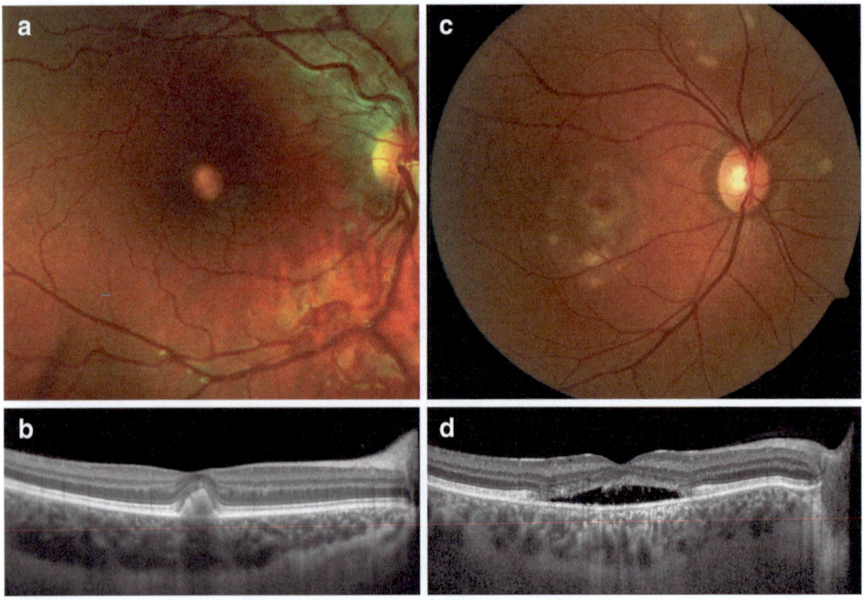

Fig. 5.12 Clinical and imaging examples of Best's disease. (**a**) Fundus photo of a 10-year-old Caucasian girl with Best disease showing a small, discrete vitelliform macular lesion. (**b**) A dense, localized hyperreflective subfoveal lesion is seen on OCT. (**c**) Fundus photo of a 46-year-old Caucasian male patient with Best disease showing fragmented vitelliform macular lesions, now at the vitelliruptive stage, with central RPE loss/atrophy. There are also satellite lesions outside the arcades superiorly and nasally to the disc. (**d**) An OCT through the top portion of the lesion shows a subretinal hyporeflective fluid pocket with marked loss of both the RPE and the ellipsoid zone (EZ) layer, despite corrected acuity of 20/50 in this eye

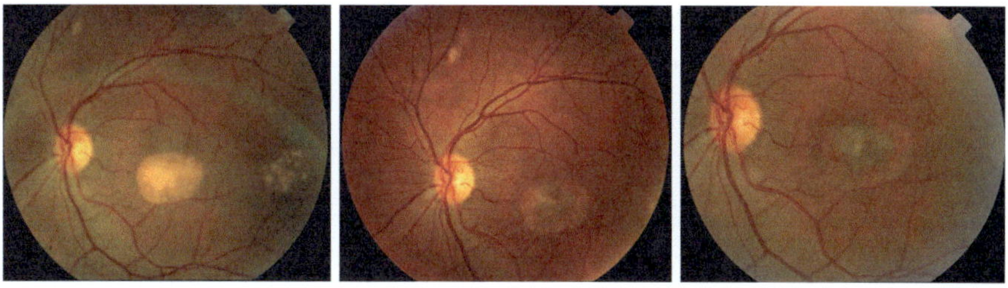

Fig. 5.13 Progression of multifocal vitelliform macular lesions. This Middle-Eastern man presented with vision loss and metamorphopsia at 41 years of age. The patient tested positive for a R218H *BEST1* mutation. In a few years, his lesions progressed through virtually all of the Best's disease stages (from right to left), reaching the atrophic stage in just a few years, leading to significant central vision loss

- *Pseudohypopyon stage*: More frequently observed around puberty. The egg yolk-like lesions disrupt and sediment inferiorly by force of gravity. The subretinal RPE appears coarsely mottled, and pigmentary clumps can develop superiorly to the pseudohypopyon. Fluid starts developing under the retina, and symptoms start occurring.
- *Vitelliruptive or "scrambled egg" stage*: Characterized by fragmentation of the yellow subretinal lesion and development of subretinal hyporeflective fluid pockets (Fig. 5.12c, d). These patients are typically symptomatic with reduced vision and metamorphopsia.
- *Atrophic stage*: The subretinal lesions disappear, leaving behind a discrete area of RPE atrophy and coarse mottling, with development of central scotomas and significant visual acuity loss.

The course of the disease can be complicated by serous or hemorrhagic macular detachment, subretinal neovascularization, disciform scarring, macular hole formation, and retinal detachment. These complications can occur virtually at any stage after the formation of the classic vitelliform lesion. Vitelliform lesions can also be seen but remain asymptomatic for years in affected individuals. Onset is usually in childhood, but it can be highly variable in some cases [27]. The progression of multifocal Best's disease lesions over several years in a patient who developed symptoms of the disease only in his early 40s is illustrated in Fig. 5.13.

Unlike classical Best's disease, ARBs often lack typical vitelliform lesions and instead appear as RP-like phenotypes with prominent exudative features and intraretinal fluid. Actual neovascular membranes have been reported in ARB patients representing more classical vitelliform subfoveal phenotypes that respond to treatment with antivascular endothelial growth factor (VEGF) [23].

Autosomal dominant vitreoretinochoroidopathy (ADViRC) is characterized by discrete bands of far peripheral abnormal chorioretinal pigmentation and atrophy, often associated with vascular (retinal vessel narrowing, occlusion, and ectasia with neovascularization) and inflammatory (vasculitis, chorioretinitis, and cystoid macula edema) changes [28, 29]. These clinical features are associated with histopatho-

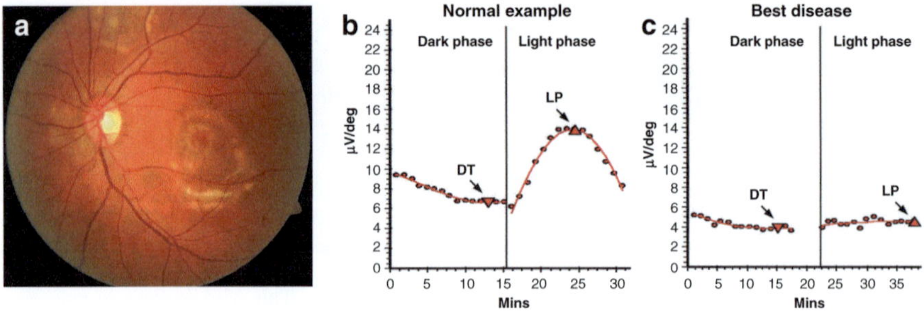

Fig. 5.14 Clinical and electrophysiological diagnostic findings for Best's disease. (**a**) Fundus photo of the left eye of the same 46-year-old Caucasian male patient with Best's disease shown in Fig. 5.12, at the vitelliruptive stage with atrophic foveal changes. Note satellite lesions superior and nasal to the disc as well. (**b**) EOG of a normal control subject. (**c**) EOG of the patient shown in (**a**) with an Arden ratio well <1.5, consistent with diagnosis of Best's disease verified molecularly to be due to an E300K *BEST1* mutation

logical and immunohistochemical evidence of inflammatory infiltration of the retina and significant VEGF upregulation [30].

The diagnosis of vitelliform macular dystrophy is mainly raised by family history (when present), fundus appearance, and fundus autofluorescence and is confirmed functionally via EOG testing [31] that by definition results in an Arden ratio below 1.5, compared to the lower limit of normal value 1.65 and, more commonly, ≥1.8 (Fig. 5.14). The mfERG may show reduced foveal or generalized peak response density mostly depending on the extent of the vitelliform lesions, the degree of macular atrophy, and depending on whether patients exhibit a multifocal vitelliform phenotype or not, resulting in reduced P1 amplitude and/or prolonged P1 implicit time in all eccentricities or central field [27] comparable to the results of OCT [32]. The ffERG is usually not informative in Best's disease except for confirming normal responses, since the extent of macular functional loss is usually insufficient to impact ffERG amplitudes. However, prior to molecular genetic diagnostic confirmation of *BEST1* gene disease-causing changes in affected patients, ffERG can be helpful to differentiate true Best's disease from phenocopies as they can be associated with *PRPH2* mutations that can also exhibit a vitelliform, more often multifocal phenotypes commonly exhibiting ffERG changes reflecting the outer segment abnormalities variably in rod and/or cone responses.

Recessive cases of pure vitelliform macular dystrophy usually produce markedly diminished Arden ratios on EOG testing and normal ffERGs, whereby they cannot be distinguished from dominant ones on the basis of clinical or functional testing alone, but only via molecular genetic diagnostic testing of affected patients and their parents, who exhibit normal EOG Arden ratios [23].

ARBs with more extensive and complex phenotypes usually exhibit not only the typical EOG findings due to bestrophin functional impairment in the RPE but also ffERG changes due to the adverse impact of the health of the overlying retina (Fig. 5.15a, b). Variably reduced rod- and cone-driven response amplitudes (from low-normal to severe) can be appear in ARBs (Fig. 5.15c). A similar pattern of ERG responses is also seen in ADViRC patients [33]. The mfERG is variably abnormal in ARBs reflecting the extent of macular involvement.

Molecular genetic diagnostic confirmation of *BEST1* gene disease-causing changes in affected patients remains paramount in these patients, including confirming parental phase of the genetic changes, confirmation that carriers of mutations do (dominant cases) or do not (recessive cases) exhibit abnormal EOG Arden ratios, and searching for larger gene changes not detected by standard sequencing techniques.

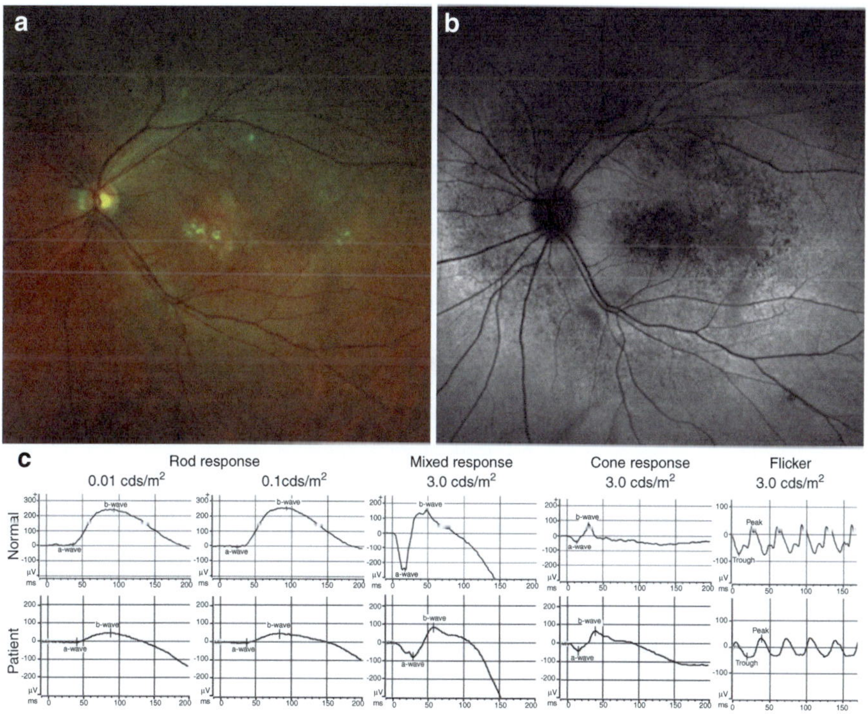

Fig. 5.15 Clinical and functional findings in autosomal recessive bestrophinopathy. (**a**) Fundus findings in a 51-year-old Caucasian female patient with molecularly confirmed autosomal recessive bestrophinopathy (ARB), due to a Tyr25Ter (Y25X) nonsense mutation, inherited paternally, and an exon 1–2 deletion of the *BEST1* gene, inherited maternally. Diffuse RPE dropout affecting the entire posterior pole with small residual vitelliform-like lesions can be seen. (**b**) The fundus autofluorescence findings in this patient show speckled hypo-autofluorescence, also outside the arcades, with some blocking effect exerted by the vitelliform-like lesions. (**c**) The ffERG of this patient shows a rod>cone pattern of retinal dysfunction with marked rod depression (dark-adapted), moderated mixed a- and b-wave reduction and delay, and normal but moderately delayed cone and flicker responses. The EOG Arden ratio in this patient was approximately 1.10 (not shown)

Pattern Dystrophies/Macular Pattern Dystrophy (MPD)

Pattern dystrophies are a group of macular dystrophies, usually inherited in autosomal dominant fashion, characterized by different "structured" patterns of yellowish subretinal deposits that usually hyper-autofluorescence on autofluorescence imaging and block fluorescence on angiography. They are presumed to contain lipofuscin, alongside RPE mottling, loss and intraretinal migration, and pigment deposition. The appearance of the patterns formed are due to the loss of the RPE and photoreceptor cells with the intact choriocapillaris and lipofuscin-containing cells in the subretinal space in the macula [34].

Pattern dystrophies usually have a good overall visual prognosis and usually exhibit late onset of symptoms compared to other macular dystrophies, such as Stargardt's disease (Fig. 5.16a–d). Because of this insidious late onset of symptoms, pattern dystrophies can often be confused also with AMD. However, due to the usually dominant pattern of inheritance, early examination of children of affected individuals usually reveals presence of initial macular signs of the disease as early as in the first decade of life (Fig. 5.16e). Slowly progressive central vision loss is possible when the deposits disrupt the integrity of the overlying ellipsoid zone and/or where there is an evolution toward macular atrophic lesions. Exudative/neovascular complications are also common and represent the main (but definitely treatable with anti-VEGF drugs) risk for significant vision loss in patients with these conditions.

The majority of the true MPDs is due to mutations in the *PRPH2* gene (also known as *peripherin/RDS*). Peripherin is expressed in the outer segments of both rods and cones and is essential to maintain the integrity of the outer segment discs. Abnormalities in the peripherin protein are known to lead to outer segment disc membrane disorganization and blebbing, and presumably, this chronic outer segment disruption is responsible for both the lipofuscin-laden deposits and the EZ loss occurring in this group of conditions. As such, despite manifestations limited to the macular region, it is not uncommon to observe abnormal ffERGs in MPD patients. Because of the greater disease emphasis on the macular region, most MPD patients with abnormal ffERGs will exhibit a cone–rod dystrophy pattern reflecting greater cone photoreceptor disease (Fig. 5.17). Accordingly, the mfERG is usually significantly abnormal in MPD patients even when the acuity may still be 20/20. It should also be remembered that protean manifestations within the same family can be present in conjunction with *PRPH2* mutations ranging from MPD, AOFVMD (see below), Stargardt-like phenotypes, cone–rod dystrophies, and retinitis pigmentosa. The type and extent of ffERG abnormalities will usually vary accordingly.

Unlike true MPDs, adult-onset foveomacular vitelliform dystrophy (AOFMVD) is a clinically and for the most part genetically distinct entity. AOFMVD usually displays small, foveal, or parafoveal elevated egg yolk-like subretinal lesions. While some patients can exhibit *PRPH2* mutations, most do not, and the genetic etiology

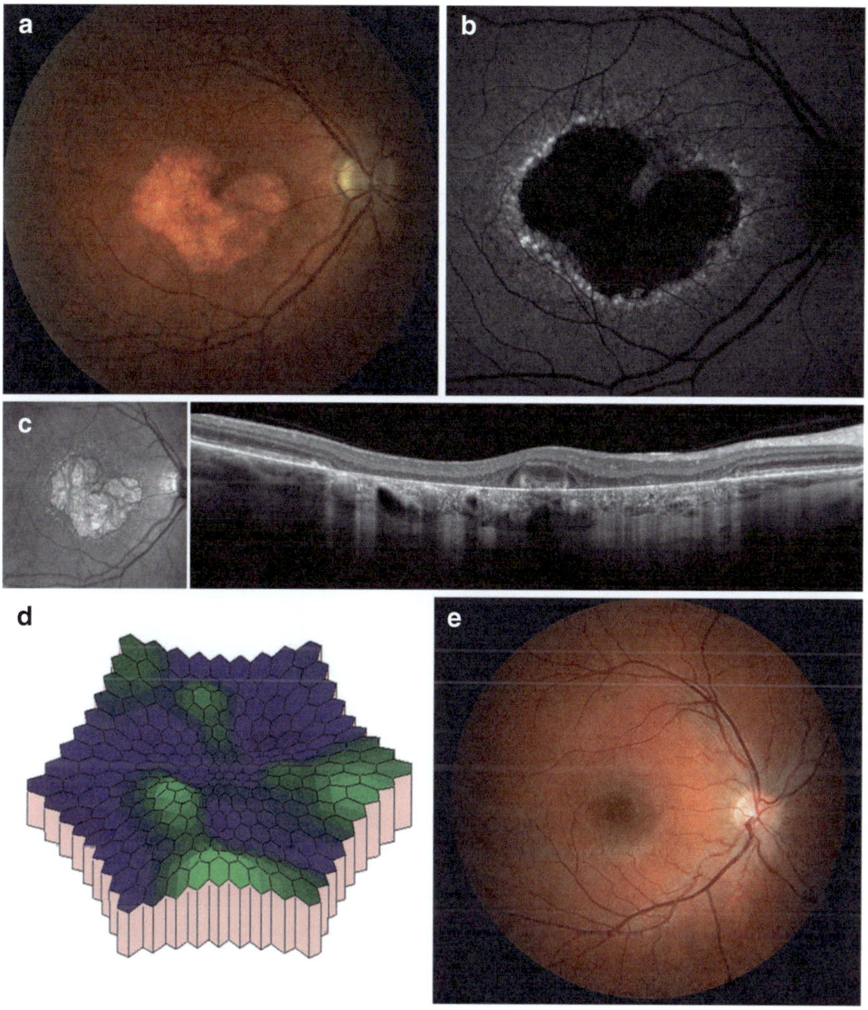

Fig. 5.16 Clinical, imaging, and functional findings in a dominant pattern dystrophy linked to a *PRPH2* mutation. (**a–d**) Fundus photos, OCT, and mfERG of a 50-year-old Caucasian man with advanced pattern dystrophy and genetically confirmed R142W pathogenic variant in the *PRPH2* gene. (**a**) Fundus photo. (**b**) fundus autofluorescence demonstrates a central area of geographic atrophy-like RPE atrophy surrounded by a thin ring of flecks. (**c**) This patient's OCT showed outer retinal atrophy with the exception of a small island at the foveola with preserved ellipsoid zone. (**d**) The mfERG of this patient shows marked response density depression throughout the tested area. (**e**) Fundus photo of the 19-year-old son of the proband, who was found to carry the same pathogenic variant in the *PRPH2* gene. While the young man is currently asymptomatic, subtle RPE mottling at the macula with a single drusenoid deposit superior to fovea can already be found. At this stage, functional testing is completely normal (not shown)

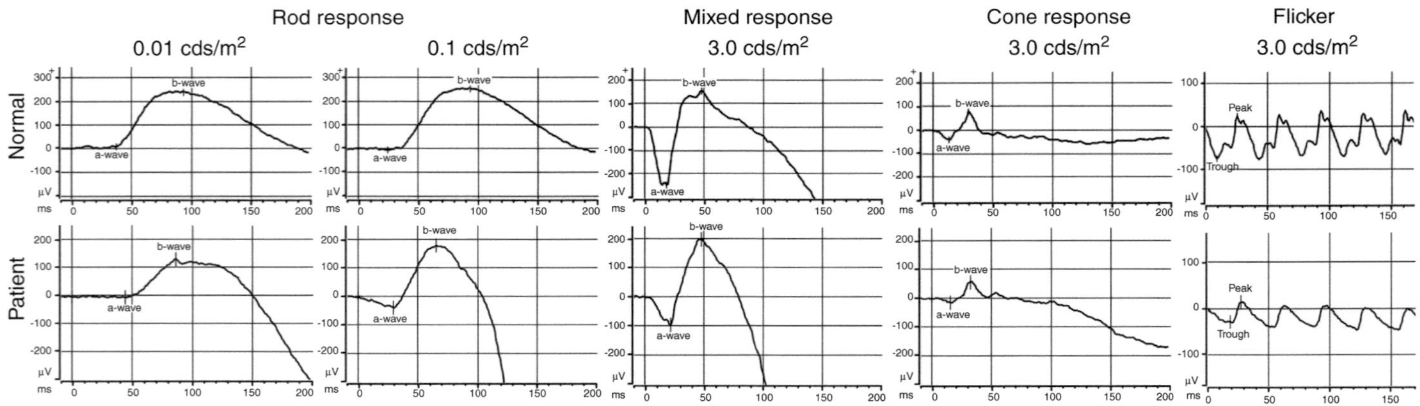

Fig. 5.17 Electroretinographic findings in a patient with a *PRPH2*-related macular pattern dystrophy. The ffERG of a patient with a *PRPH2* disease-causing variant showing mildly attenuated rod-driven responses, reduced mixed a-wave, borderline low mixed b-wave, mildly reduced transient cone responses, and even more reduced flicker responses, consistent with a mild cone–rod dystrophy pattern

of AOFVMD remains often undetermined. However, AOFVMD-type patients are also possible in bona fide *BEST1*-related macular dystrophies. In the latter subgroup of diseases, a dominant history often exists, and the EOG test is usually abnormal (<1.5 Arden ratio) or borderline (1.5–1.65 range). However, in the remainder of the cases, the EOG is usually at least in the low normal range (>1.7 Arden ratio). Unless the causal gene is *PRPH2*, the ffERG is usually normal or minimally abnormal in AOFMVD patients, whereas the mfERG is usually reduced in variable fashion.

Doyne's Honeycomb Macular Dystrophy (Malattia Leventinese)

Doyne honeycomb macular dystrophy (DHMD), also known as Malattia Leventinese, Radial Drusen, or Dominant Drusen, is characterized by the drusen accumulating under RPE and a complaint of early-onset dark adaptation problems. Over time, drusen grow in extent and connect together to form either linear or honeycomb reticular patterns (Fig. 5.18). The condition is inherited in a dominant fashion, and, to date, a single *EFEMP1* gene mutation has been identified in all cases of DHMD [35].

The onset of DHRD is usually in or after 30–40 years of age. In the early stage of DHRD, symptoms may include decreased visual acuity, metamorphopsia, abnormal color vision, photophobia, and relative scotomas. In the later stage of DHRD usually occurring by 40–50 years of age, the central vision is further decreased with absolute scotomas. If choroidal neovascularization (CNV) occurs, the vision loss can develop faster and further. In DHRD, the amplitude of pattern ERG is usually decreased while ffERG rod and rod–cone responses and EOG Arden ratio are usually not affected [36].

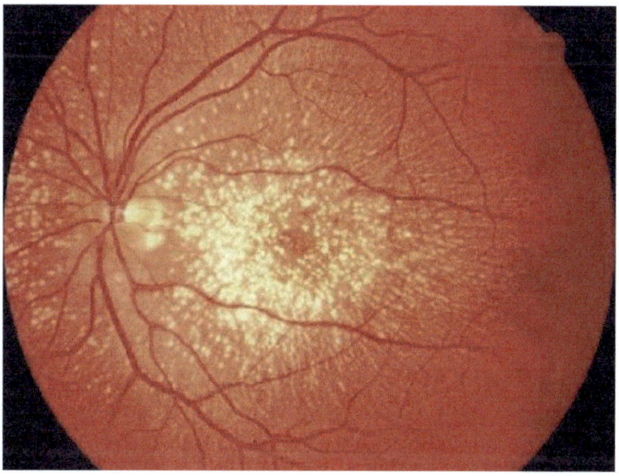

Fig. 5.18 Fundus findings in Doyne's honeycomb macular dystrophy (Malattia Leventinese). Fundus photo of a patient with DHMD showing confluent yellow drusen in a streak-like radial configuration characteristic of the entity (Adapted from Figure 1 [35])

Occult Macular Dystrophy

Occult macular dystrophy (OMD) is an autosomal dominant disease caused by heterozygous mutations in the *RP1L1* gene [37]. Its symptoms are highly variable. Its vision progressively declines, starting from adolescence to the sixth decade of life or even older. Even those who complain of changes in acuity may still have 20/20 to 20/25 vision, but it may also be much worse, in the range of count fingers. There can be considerable asymmetry in acuity between the two eyes, but there is little known about the rate of vision loss. Mild dyschromatopsia is often observed with deficits in both red-green and red-green discrimination, but total color blindness has also been reported.

The fundus appearance and fluorescein angiography of OMD do not typically show clinically visible abnormalities even in advanced stages. The ffERG is usually normal in OMD or at the most slightly decreased [38–40]. However, the mfERG is typically significantly decreased in OMD patients (Fig. 5.19). These findings are usually accompanied by visible changes by OCT imaging, that can include a range of findings, such as exudative complications with subretinal fluid resembling central serous chorioretinopathy or Best's disease [40–43]. The more common OCT finding is a fine nonspecific fragmentation of the ellipsoid zone [43], as it can also be seen in many other macular and cone dystrophies. Specialized recordings outside of the clinical standards measuring ON and OFF focal macular responses with long-duration stimuli have shown defects in these responses [44], and focal flash ERGs are also abnormal in OMD [45]. The diagnosis of certainty of OMD can ultimately be confirmed only via molecular genetic diagnostic testing. It is worth noting mutations in the *RP1L1* gene can cause also autosomal recessive retinitis pigmentosa.

North Carolina Macular Dystrophy (NCMD)

NCMD, also known as central areolar pigment epithelial dystrophy (CAPED) [46], was first described by Lefler et al. in a group of Irish descendants living in the North Carolina mountains [47], and it was the first macular dystrophy mapped reliably (*MCDR1* locus) [48]. Even though it was named after a family in NC, the dystrophy has been diagnosed in families all over the world, mapping to the same *MCDR1* locus [46, 49–51]. It is inherited in an autosomal dominant fashion with high penetrance [46, 51].

Its phenotypic expression consists of three grades: (1) fine macular drusen; (2) confluent drusen, choroidal neovascularization, or disciform scar; and (3) macular staphylomas, or coloboma- or toxoplasmosis-like chorioretinal excavations (Fig. 5.20) [48, 51, 52]. The ffERG and EOG are typically normal in NCMD [49, 50], whereas the pattern ERG [50] and the mfERG will be reduced in grade 3 disease, whereas they will typically remain normal in the earlier stages of the disease

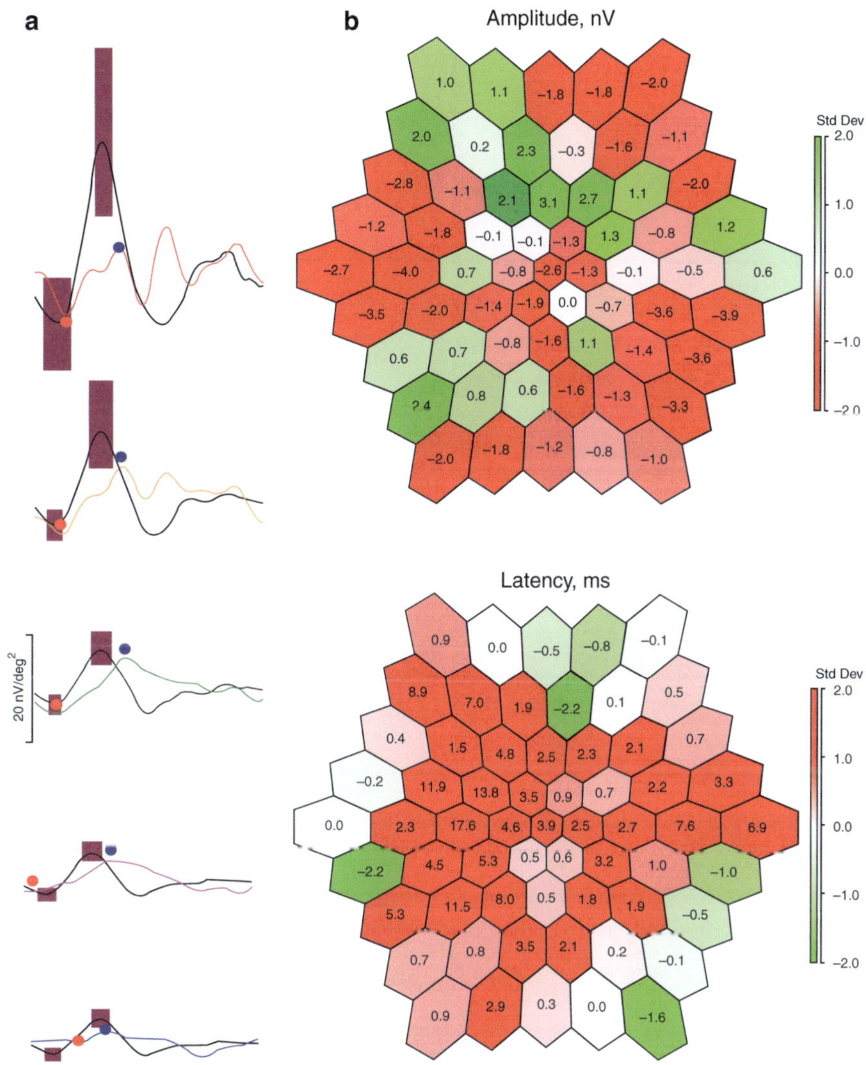

Fig. 5.19 Multifocal ERG findings in occult macular dystrophy (OMD). The mfERG of a patient with a disease-causing mutation (Glu1466Lys or E1466K) in the *RP1L1* gene and with an observed occult macular dystrophy phenotype. The (**a**) average of the P1 trace arrays by eccentricity rings (the purple bars indicate 95% confidence intervals of normal subjects) and (**b**) the amplitude and latency enface plots standard deviation plots from the mean are shown, to illustrate the patchy response reduction and the generalized delays thereof, more severe centrally and pericentrally

unless there are neovascular complications. All these conditions were met in the cases illustrated in Fig. 5.20. Of note, cases who exhibit only a small patch of flat RPE loss (of the central areolar pigment epithelial dystrophy or CAPED type) away from the fovea region that can be completely asymptomatic have also been observed in NCMD families.

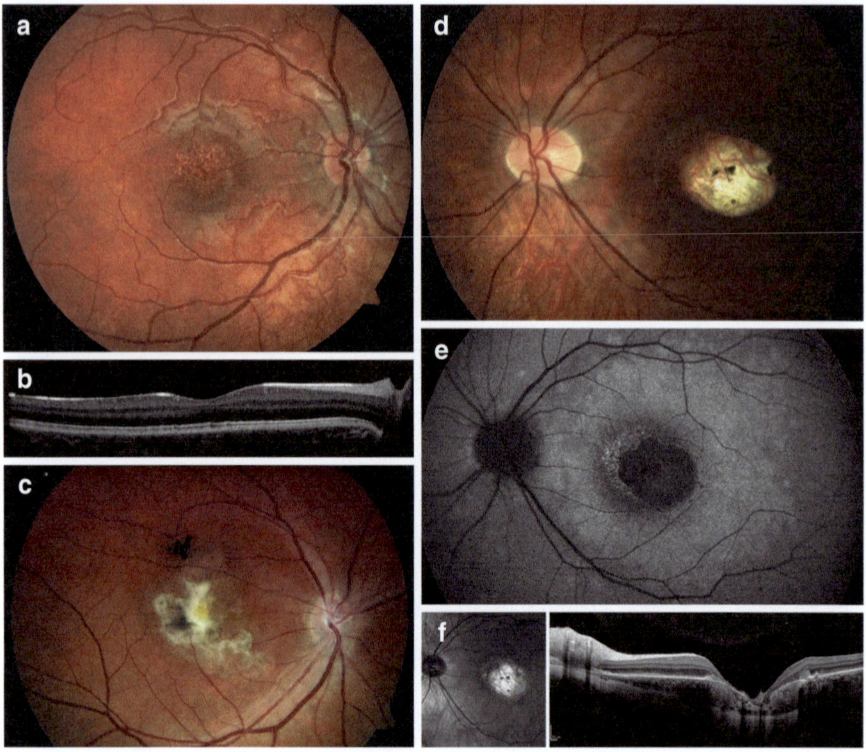

Fig. 5.20 Phenotypic expression of North Carolina macular dystrophy (NCMD). (**a**) Fundus photo of an asymptomatic 4-year-old Caucasian girl noted to have macular changes by her optometrist, illustrating grade 1 NCMD. (**b**) The macular OCT of this same patient is shown, revealing only microdrusen-like changes in the foveal region. (**c**) Her 32-year-old father exhibited a grade 2 case of NCMD with a fibrotic subfoveal neovascular complication in the right eye. The left eye showed only confluent drusen (not shown). Remarkably, the retina and the RPE overlying the fibrotic scar were both intact, whereby acuity also in the right eye was 20/25. The ffERG of both patients and the EOG of the father were completely normal (not shown). The father's mfERG was mildly abnormal only in the right eye (not shown). (**d**) Another case of a Caucasian woman in her 40s with lifelong history of reduced acuity, central scotoma, and macular lesions, exhibiting a grade 3 NCMD with a macular toxoplasmosis-like coloboma. (**e**) On fundus autofluorescence, there is marked central hypo-autofluorescence corresponding to the chorioretinal coloboma, with a surrounding ring of discrete brighter hyper-autofluorescence on the nasal side and a fainter halo of hyper-autofluorescence around the excavated lesion. (**f**) On OCT, the chorioretinal excavation is clearly delineated, with loss of retinal, RPE, and choroidal tissue. This patient had a son with a very similar lesion, whereas her mother exhibited only a small patch of flat RPE loss (of the central areolar pigment epithelial dystrophy or CAPED type) away from the fovea region and was completely asymptomatic

References

1. Bax NM. The portal for rare diseases and orphan drugs. Orphanet: About Orphan Drugs. 2019. Available from: www.orpha.net/consor/cgi-bin/OC_Exp.php?Lng=GB&Expert=827 http://www.orpha.net/consor/cgi-bin/OC_Exp.php?Lng=GB&Expert=827.
2. Allikmets R, et al. Mutation of the Stargardt disease gene (ABCR) in age-related macular degeneration. Science. 1997;277(5333):1805–7.
3. Michaelides M, Hunt DM, Moore AT. The genetics of inherited macular dystrophies. J Med Genet. 2003;40(9):641–50.
4. Boon CJ, et al. Mutations in the peripherin/RDS gene are an important cause of multifocal pattern dystrophy simulating STGD1/fundus flavimaculatus. Br J Ophthalmol. 2007;91(11):1504–11.
5. Cideciyan AV, et al. Mutations in ABCA4 result in accumulation of lipofuscin before slowing of the retinoid cycle: a reappraisal of the human disease sequence. Hum Mol Genet. 2004;13(5):525–34.
6. Fishman GA, et al. Variation of clinical expression in patients with Stargardt dystrophy and sequence variations in the ABCR gene. Arch Ophthalmol. 1999;117(4):504–10.
7. Kuniyoshi K, et al. Multifocal electroretinograms in Stargardt's disease/fundus flavimaculatus. Ophthalmologica. 2014;232(2):118–25.
8. Haas J. Ueber das zusammenvorkommen von Veranderungen der retina und choroidea. Arch Augenheilkd. 1898;37:343–8.
9. George ND, Yates JR, Moore AT. X linked retinoschisis. Br J Ophthalmol. 1995;79(7):697–702.
10. Wu WW, Molday RS. Defective discoidin domain structure, subunit assembly, and endoplasmic reticulum processing of retinoschisin are primary mechanisms responsible for X-linked retinoschisis. J Biol Chem. 2003;278(30):28139–46.
11. Sieving PA, Yashar BM, Ayyagari R. Juvenile retinoschisis: a model for molecular diagnostic testing of X-linked ophthalmic disease. Trans Am Ophthalmol Soc. 1999;97:451–64; discussion 464–9.
12. Cukras CA, et al. Analysis of anatomic and functional measures in X-linked retinoschisis. Invest Ophthalmol Vis Sci. 2018;59(7):2841–7.
13. Sieving PA, MacDonald IM, Chan S. X-linked juvenile retinoschisis. In: Adam MP, et al., editors. GeneReviews(R). University of Washington, Seattle; 2014.
14. Iannaccone A. Optical coherence tomography in rare pediatric cases. Retina Today. 2012(September). pp. 1–3.
15. Iannaccone A, et al. An unusual X-linked retinoschisis phenotype and biochemical characterization of the W112C RS1 mutation. Vision Res. 2006;46(22):3845–52.
16. Ghajarnia M, Gorin MB. Acetazolamide in the treatment of X-linked retinoschisis maculopathy. Arch Ophthalmol. 2007;125(4):571–3.
17. Miyake Y, et al. Focal macular electroretinogram in X-linked congenital retinoschisis. Invest Ophthalmol Vis Sci. 1993;34(3):512–5.
18. Kim LS, et al. Multifocal ERG findings in carriers of X-linked retinoschisis. Doc Ophthalmol. 2007;114(1):21–6.
19. Huang S, et al. The multifocal electroretinogram in X-linked juvenile retinoschisis. Doc Ophthalmol. 2003;106(3):251–5.
20. White K, Marquardt A, Weber BH. VMD2 mutations in vitelliform macular dystrophy (Best disease) and other maculopathies. Hum Mutat. 2000;15(4):301–8.

21. Marchant D, et al. New VMD2 gene mutations identified in patients affected by Best vitelliform macular dystrophy. J Med Genet. 2007;44(3):e70.
22. Boon CJ, et al. The spectrum of ocular phenotypes caused by mutations in the BEST1 gene. Prog Retin Eye Res. 2009;28(3):187–205.
23. Iannaccone A, et al. Autosomal recessive best vitelliform macular dystrophy: report of a family and management of early-onset neovascular complications. Arch Ophthalmol. 2011;129(2):211–7.
24. Burgess R, et al. Biallelic mutation of BEST1 causes a distinct retinopathy in humans. Am J Hum Genet. 2008;82(1):19–31.
25. Boon CJ, et al. Autosomal recessive bestrophinopathy: differential diagnosis and treatment options. Ophthalmology. 2013;120(4):809–20.
26. Davidson AE, et al. Missense mutations in a retinal pigment epithelium protein, bestrophin-1, cause retinitis pigmentosa. Am J Hum Genet. 2009;85(5):581–92.
27. Renner AB, et al. Late onset is common in best macular dystrophy associated with VMD2 gene mutations. Ophthalmology. 2005;112(4):586–92.
28. Kaufman SJ, et al. Autosomal dominant vitreoretinochoroidopathy. Arch Ophthalmol. 1982;100(2):272–8.
29. Blair NP, et al. Autosomal dominant vitreoretinochoroidopathy (ADVIRC). Br J Ophthalmol. 1984;68(1):2–9.
30. Goldberg MF, et al. Ocular histopathology and immunohistochemical analysis in the oldest known individual with autosomal dominant vitreoretinochoroidopathy. Ophthalmol Retina. 2018;2(4):360–78.
31. MacDonald IM, Lee T. Best vitelliform macular dystrophy. In: Pagon RA, et al., editors. GeneReviews(R). University of Washington, Seattle; 1993.
32. Glybina IV, Frank RN. Localization of multifocal electroretinogram abnormalities to the lesion site: findings in a family with Best disease. Arch Ophthalmol. 2006;124(11):1593–600.
33. Lafaut BA, et al. Clinical and electrophysiological findings in autosomal dominant vitreoretinochoroidopathy: report of a new pedigree. Graefes Arch Clin Exp Ophthalmol. 2001;239(8):575–82.
34. Zhang K, et al. Butterfly-shaped pattern dystrophy: a genetic, clinical, and histopathological report. Arch Ophthalmol. 2002;120(4):485–90.
35. Stone EM, et al. A single EFEMP1 mutation associated with both Malattia Leventinese and Doyne honeycomb retinal dystrophy. Nat Genet. 1999;22(2):199–202.
36. Haimovici R, et al. Symptomatic abnormalities of dark adaptation in patients with EFEMP1 retinal dystrophy (Malattia Leventinese/Doyne honeycomb retinal dystrophy). Eye (Lond). 2002;16(1):7–15.
37. Akahori M, et al. Dominant mutations in RP1L1 are responsible for occult macular dystrophy. Am J Hum Genet. 2010;87(3):424–9.
38. Miyake Y, Tsunoda K. Occult macular dystrophy. Jpn J Ophthalmol. 2015;59(2):71–80.
39. Hayashi T, et al. Autosomal dominant occult macular dystrophy with an RP1L1 mutation (R45W). Optom Vis Sci. 2012;89(5):684–91.
40. Kondo M, et al. Occult macular dystrophy in an 11 year old boy. Br J Ophthalmol. 2004;88(12):1602–3.
41. Piao CH, et al. Multifocal electroretinogram in occult macular dystrophy. Invest Ophthalmol Vis Sci. 2000;41(2):513–7.
42. Ahn SJ, et al. Multimodal imaging of occult macular dystrophy. JAMA Ophthalmol. 2013;131(7):880–90.
43. Tsunoda K, et al. Clinical characteristics of occult macular dystrophy in family with mutation of RP1l1 gene. Retina. 2012;32(6):1135–47.
44. Kondo M, Miyake Y. Assessment of local cone on- and off-pathway function using multifocal ERG technique. Doc Ophthalmol. 2000;100(2–3):139–54.

45. Kabuto T, et al. A new mutation in the RP1L1 gene in a patient with occult macular dystrophy associated with a depolarizing pattern of focal macular electroretinograms. Mol Vis. 2012;18:1031–9.
46. Small KW, et al. North Carolina macular dystrophy and central areolar pigment epithelial dystrophy. One family, one disease. Arch Ophthalmol. 1992;110(4):515–8.
47. Lefler WH, Wadsworth JA, Sidbury JB Jr. Hereditary macular degeneration and aminoaciduria. Am J Ophthalmol. 1971;71(1 Pt 2):224–30.
48. Small KW, et al. North Carolina macular dystrophy phenotype in France maps to the MCDR1 locus. Mol Vis. 1997;3:1.
49. Sauer CG, et al. An ancestral core haplotype defines the critical region harbouring the North Carolina macular dystrophy gene (MCDR1). J Med Genet. 1997;34(12):961–6.
50. Reichel MB, et al. Phenotype of a British North Carolina macular dystrophy family linked to chromosome 6q. Br J Ophthalmol. 1998;82(10):1162–8.
51. Yang Z, et al. Clinical characterization and genetic mapping of North Carolina macular dystrophy. Vision Res. 2008;48(3):470–7.
52. Khurana RN, et al. A reappraisal of the clinical spectrum of North Carolina macular dystrophy. Ophthalmology. 2009;116(10):1976–83.

Chapter 6
Degenerative Night-Blinding Disorders and Cone and Cone–Rod Dystrophies

Wajiha Jurdi Kheir, Minzhong Yu, Alfonso Senatore, Alessandro Racioppi, Roberto Gattegna, Donnell Creel, and Alessandro Iannaccone

Retinitis Pigmentosa

Retinitis pigmentosa (RP), also called rod–cone dystrophy, is a group of inherited retinal diseases characterized by the progressive degeneration of rod photoreceptor cells followed by the disorder or death of cone photoreceptors and retinal pigment epithelium (Fig. 6.1). It affects approximately 1:3,500–1:4,000 people. Mutations in over 130 genes have been identified to be related to non-syndromic or syndromic RP [1, 2]. In most cases, the RP symptoms only involve loss of vision, classified as non-syndromic RP. The inherited patterns of non-syndromic RP include autosomal recessive (50–60%), autosomal dominant (30–40%), and X-linked (5–15%) [1, 2]. In general, autosomal dominant RP tends to be the slowest and most variable in expression, while X-linked RP tends to have consistently the earliest onset. In

W. J. Kheir · A. Senatore · A. Iannaccone
Center for Retinal Degenerations and Ophthalmic Genetic Diseases, Duke University School of Medicine, Duke Eye Center, Department of Ophthalmology, Durham, NC, USA

M. Yu (✉)
Department of Ophthalmology, University Hospitals Eye Institute, Cleveland, OH, USA
e-mail: minzhong.yu@uhhospitals.org

A. Racioppi
Center for Retinal Degenerations and Ophthalmic Genetic Diseases, Duke University School of Medicine, Duke Eye Center, Department of Ophthalmology, Durham, NC, USA

University of North Carolina, Chapel Hill, NC, USA

R. Gattegna
Center for Retinal Degenerations and Ophthalmic Genetic Diseases, Duke University School of Medicine, Duke Eye Center, Department of Ophthalmology, Durham, NC, USA

Retina Service, Israelitic Hospital, Rome, Italy

D. Creel
Moran Eye Center, University of Utah School of Medicine, Salt Lake City, UT, USA

M. Yu et al. (eds.), *Handbook of Clinical Electrophysiology of Vision*,
https://doi.org/10.1007/978-3-030-30417-1_6

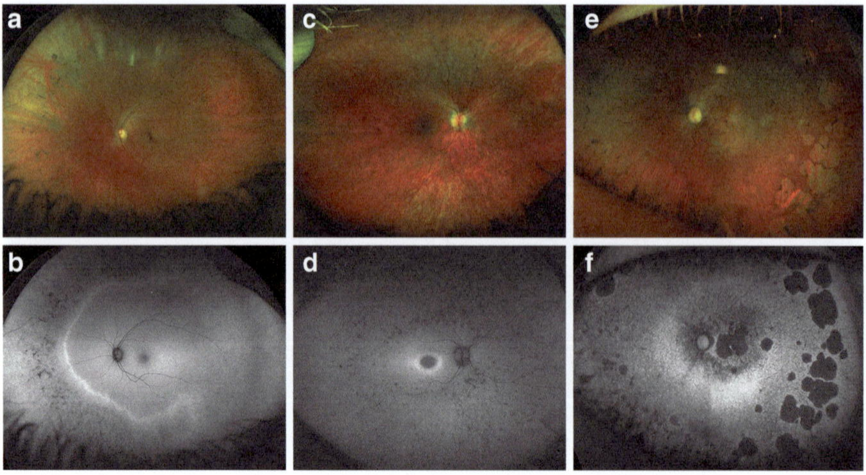

Fig. 6.1 Imaging findings in retinitis pigmentosa. Examples of fundus photos and fundus autofluorescence (FAF) in three patients all molecularly confirmed to have RP. (**a**) Patient 1: Early-stage autosomal recessive RP, with bone spicule-like deposits and RPE, changes predominantly in the peripheral nasal hemiretina. The macula appears normal, but the retinal vasculature is already attenuated, and there is waxy disc pallor. (**b**) A crescent-shaped ring of hyper-autofluorescence delineates the leading edge of the peripheral disease of Patient 1. There is speckled hypo-autofluorescence nasally in the area of bone spicule-like pigment deposition and mottled RPE. (**c**) Patient 2 has intermediate stage autosomal dominant RP. Fundus examination shows fine punched-out lesions (moth-eaten appearance) and bone spicule-like peripheral pigment deposits, attenuated vessels, questionable disc pallor, and attenuated foveal reflexes. (**d**) FAF reveals a ring of hyper-autofluorescence around the fovea demarcating the leading edge of the disease (good ellipsoid zone preservation on OCT is found within these boundaries, not shown). Speckled hypo-autofluorescence nasally in the area of bone spicule-like pigment deposition and moth-eaten RPE. (**e**) Patient 3 has advanced autosomal recessive RP with atrophic macular involvement. This patient shows large areas of nummular confluent RPE atrophy and bone spicule pigment deposits in the periphery, attenuated vessels, peripapillary atrophy, waxy disc pallor, and macular atrophy. (**f**) On FAF, there is severe hypo-autofluorescence throughout the posterior pole and in the large peripheral nummular patches of RPE atrophy beyond the arcade. A large ring of speckled hyper-autofluorescence around the arcades and a large ring of speckled hypo-autofluorescence around the central atrophic areas can be seen, indicative of widespread RPE suffering

many cases, RP occurs as one of the syndromes involving other disorders in the body. These syndromes are classified as syndromic RP, including Usher syndrome, Bardet–Biedl syndrome, Refsum disease, and the neuropathy, ataxia, and retinitis pigmentosa (NARP) syndrome. These syndromes will be discussed in other dedicated sections. All types of RP generally cause night blindness in the early stage, followed by reduction of peripheral visual field, and ultimately legal blindness. The development of RP usually occurs over several decades. Therefore, early diagnosis and treatment can reduce the rate of RP progression and extend the persistence of useful visual function for additional years [1].

There is no marked difference in the characteristics of electrophysiologic tests in different types of RP except the time frame of development. For the diagnosis of RP, visual field testing, ffERGs, spectral domain optical coherence tomography (SD-OCT), fundus autofluorescence (FAF), and at times fluorescein angiography are the cornerstones of the clinical and functional workup of symptomatic patients. Typically, the ffERG exhibits measurable reduction or other qualitative changes even at the earliest stages of RP when the symptoms are still mild, although if only small areas of the retina are affected early on, SD-OCT and FAF can detect the earliest signs of localized diseases better than the ffERG. Rod function is usually decreased first in RP, preceding the reduction of cone function, or the two can occur concomitantly, as it is often the case in certain forms of dominant RP (e.g., the ones linked to *PRPH2* mutations) as well as in X-linked RP, in which both rods and cones are affected early and at times equally [1]. In more advanced disease stages, the ffERG of RP becomes non-recordable or near the noise threshold (Fig. 6.2).

The mapping of cone photoreceptor function across the macular region can be tested by mfERG usually showing a decrease in amplitude starting from the peripheral field and spreading toward the central field as the RP progresses. Response of mfERG is still recordable in some cases even when the ffERG has become undetectable [3].

Leber's Congenital Amaurosis/EORP/SECORD

Leber's congenital amaurosis (LCA) is a type of congenital or early-onset inherited retinal dystrophy within the first 6 months of life with severe ocular symptoms, such as extremely low visual acuity in infancy, nystagmus, impaired pupillary-light reflex, and normal fundus appearance at the early stage followed by gradually pigmentary change in later stages. If these symptoms are mild and present after 1 year of age, the disorder is often called early-onset RP (EORP) or severe early childhood onset retinal degeneration (SECORD) [4]. LCA can be caused by the mutations of more than 16 genes, including *AIPL1*, *CEP290*, *CRB1*, *CRX*, *GUCY2D*, *IMPDH1*, *IQCB1*, *LCA5*, *LRAT*, *MERTK*, *RD3*, *RDH12*, *RPGRIP1*, *RPE65*, *SPATA7*, and *TULP1* (Fig. 6.3). LCA is generally inherited in an autosomal recessive pattern, but mutations in *CRX* and *IMPDH1* can cause autosomal dominant LCA [5, 6]. The ffERG in LCA typically show non-recordable or severely reduced responses very early on [7–11]. However, cases where the ffERG remains however partially recordable into the second and even third decade of life are seen (Fig. 6.4).

Enhanced S-Cone Syndrome

Enhanced S-cone syndrome (ESCS), also as known as Goldmann–Favre vitreoretinal dystrophy or clumped pigmentary retinal dystrophy (CPRD), is a rare inherited autosomal recessive night-blinding retinal dystrophy [12, 13]. ESCS is caused by mutations in the *NR2E3* gene, encoding a ligand-dependent transcription factor for retinal progenitor cell fate producing the development of retinas with virtually no

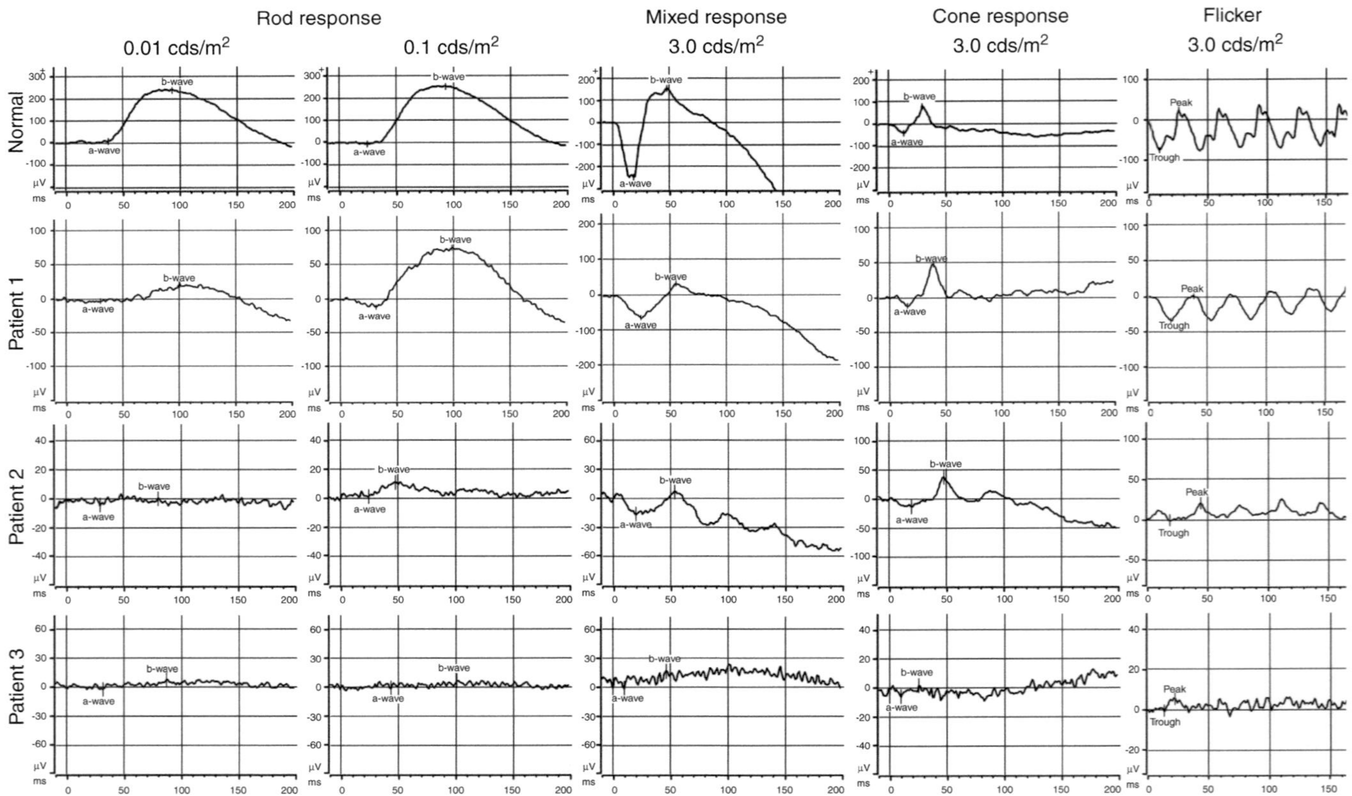

Fig. 6.2 Electroretinographic findings in retinitis pigmentosa. The ERGs of Patient 1, 2, and 3 are shown. Patient 1 has moderately reduced rod/mixed responses and normal to mildly reduced and delayed cone responses, consistent with early-stage RP. Patient 2 has severely reduced rod responses, markedly reduced mixed responses, and moderately reduced/delayed cone responses consistent with intermediate stage RP. Patient 3 has non-recordable rod, mixed, and cone responses consistent with severe RP

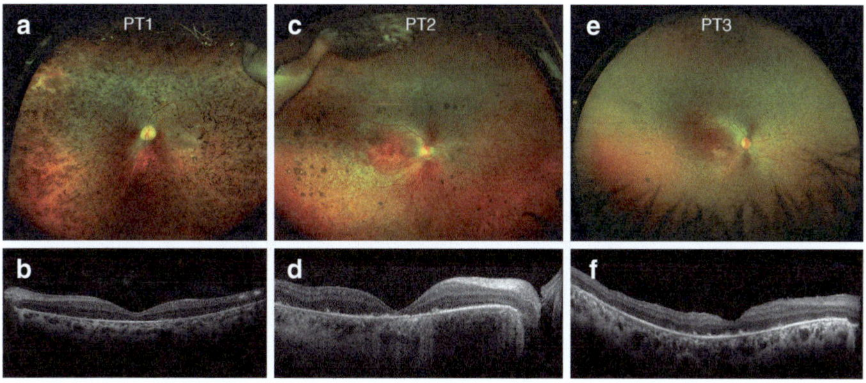

Fig. 6.3 Clinical and imaging findings in LCA/EORP/SECORD patients. All patients displayed congenital (LCA) or early childhood-onset (EORP/SECORD) nystagmus and evidence of poor vision and night blindness. (**a**) PT1: Fundus photo of a 19-year-old Caucasian female with LCA due to a deletion leading up to a frameshift effect in the *RDH12* gene. Note the widespread pigmentary deposits with marked vascular attenuation and only partial macular sparing. There is partial paravascular RPE sparing. (**b**) The OCT of this patient shows essentially no detectable EZ or ELM residue across the entire macular region despite otherwise good preservation of all other layers. (**c**) PT2: Fundus photo of a 25-year-old African-American male with LCA due to compound heterozygosity for a missense and a nonsense mutation in the *RPE65* gene. The retinal degeneration of this patient shows disseminated nummular pigment deposits and widespread RPE dropout, including atrophy at the macular level. (**d**) The OCT of PT2 shows severe ellipsoid zone (EZ)/External limiting membrane (ELM) loss across the macula with hyper-transmission defects due to the associated RPE atrophy. The inner retina appears to be abnormally thickened, especially at the retinal nerve fiber layer (RNFL) level. (**e**) PT3: Fundus photo of a 26-year-old African-American female patient with EORP/SECORD, also due to compound heterozygosity for a missense and a nonsense mutation in the *RPE65* gene (different than PT2). The retinal degeneration of this patient shows only a few nummular pigment deposits and widespread RPE dropout, including mild bull's eye-shaped atrophy at the macular level. (**f**) The OCT of PT3 shows severe EZ/ELM loss across the macula with seemingly good ONL residue and without hyper-transmission defects. The inner retina appears to be well preserved, with questionable thickening of the RNFL

rods and dysplastic retinas with an abnormally large number of blue light-sensitive (S)-cones in place of the rods [14–17]. For this reason, affected patients usually experience congenital night blindness. The visual acuities and visual fields (typically exhibiting a ring scotoma) of ESCS patients tend to worsen slowly with age [18], although central acuity can be threatened early and become markedly reduced as the result of macular retinoschisis typically present in these patients, can develop both insidiously (more typically) and even acutely, and is typically responsive to carbonic anhydrase inhibitors [12, 19, 20].

The typical fundus changes of ESCS (Fig. 6.5) occur primarily at the vascular arcades, where normally the highest rod cell density would be, and differ fundamentally from RP, showing deep, nummular, confluent pigment deposits, and patches of atrophy underneath the retina or in the outer retinal layers, unlike the bone spicule-like deposits characterizing RP and the track the inner retinal blood vessels and capillaries [12]. In more advanced stages, the nummular confluent deposits are usually pigmented, hence the CPRD definition, and have been known to resemble laser

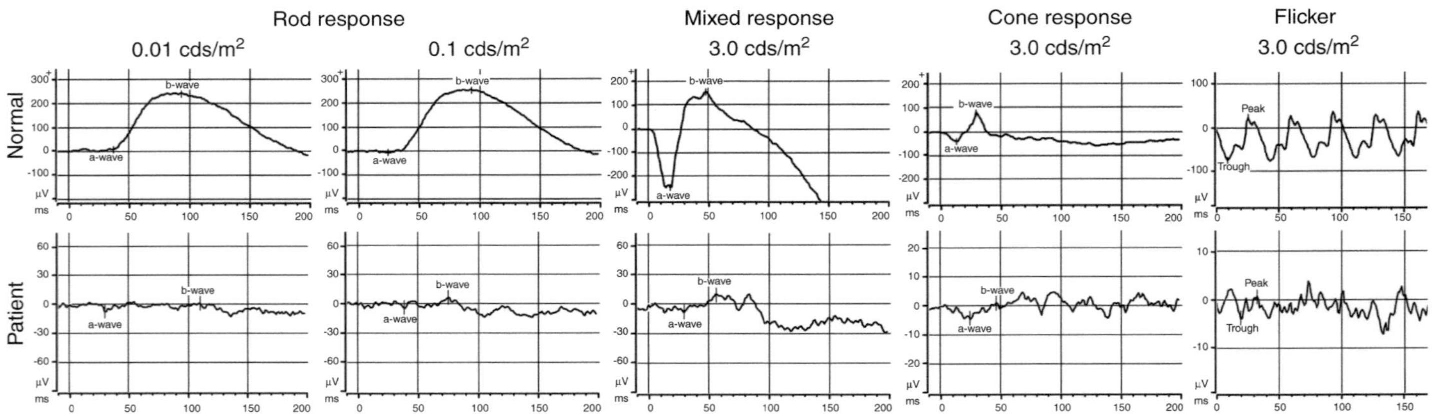

Fig. 6.4 Example of ffERG in LCA patient with recordable responses. The ffERG of PT1 in Fig. 6.3 (homozygous for a deletion in the *RDH12* gene) shows non-recordable rod responses and severely reduced yet clearly measurable mixed and cone responses consistent with advanced rod–cone pattern of disease. Despite the apparent severity of the disease, even in a patient meeting the diagnostic criteria for LCA, the ffERG can remain detectable into the second and even third decade of life. A younger patient also exhibiting *RDH12* mutations showed even more robust responses at the age of 7 years old (not shown)

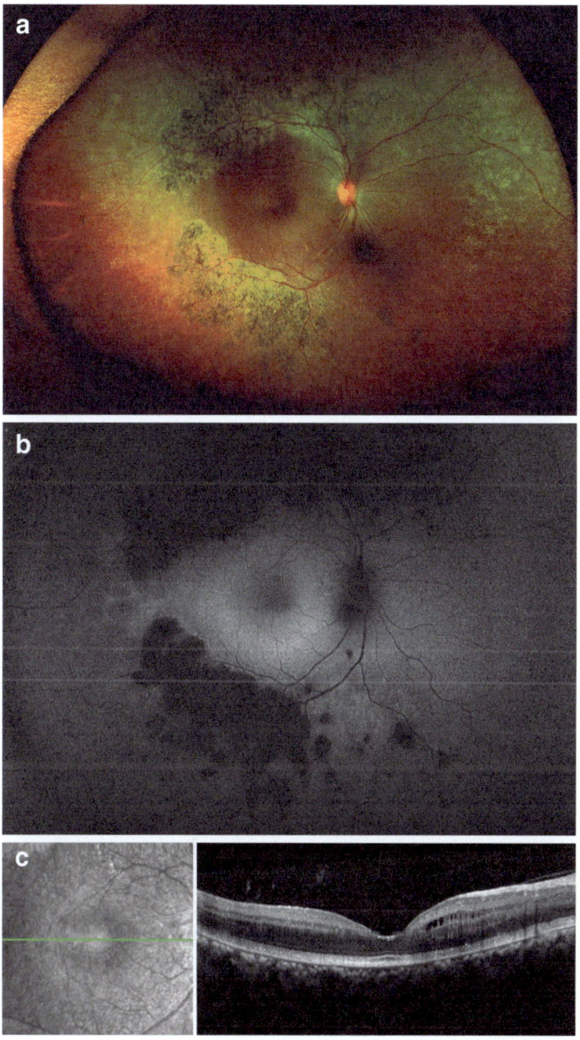

Fig. 6.5 Clinical and imaging findings in enhanced S-cone syndrome (ESCS). Findings in a 14-year-old Ethiopian girl affected with congenital night blindness due to ESCS resulting from compound heterozygous mutations in the *NR2E3* gene. (**a**) Fundus photography shows superotemporal and inferotemporal deep nummular pigment deposits at the arcades associated with RPE atrophy and 360-degree micro-exudates and flecks in the periphery. (**b**) Fundus autofluorescence shows hypo-fluorescence corresponding to the pigment deposits and the RPE loss seen clinically and a faint ring of hyper-fluorescence at and inside the arcades. (**c**) OCT shows a grainy EZ and mild microcystic parafoveal schitic changes without significant retinal thinning or EZ, ELM, or ONL loss. Consistent with this finding, acuity in this eye is approximately 20/20

burns. However, in the earliest stages of ESCS, these deep deposits at the arcades can appear whitish, in the guise of flecks, and are often associated with punctate more superficial changes, as in the case shown in Fig. 6.5. Furthermore, the retinal blood vessels of ESCS patients are rarely attenuated, except in the very late stages of the diseases, and also the typical waxy pallor of the RP disc is customarily absent in ESCS patients (Fig. 6.5). Lastly, unlike most patients with RP, ESCS patients are not only prone to macular retinoschisis but also to peripheral schisis, and this can lead to rhegmatogenous retinal tears and detachments posing a potentially significant and acute threat to the peripheral vision of ESCS patients [12]. The more whitish superficial deposits patients with ESCS develop are often associated with subclinical levels of peripheral schisis.

The unusually large number of S-cones in ESCS causes, in its most typical appearance, non-recordable rod responses and supernormal mixed and cone ffERG transient responses typically associated with a distinctive electronegative delayed waveform that persists in both characteristics also under light-adapted conditions [21–23]. In many cases, however, both the mixed and the cone-driven ffERG responses are reduced in amplitude, albeit maintaining the distinctive electronegative delayed waveform typical for ESCS (Fig. 6.6). Pattern ERG P50 peaks are generally delayed and/or decreased significantly. Multifocal ERG shows decreased response densities in the entire tested field or in peripheral field [20, 24, 25].

Choroideremia

Choroideremia is a rare X-linked recessive retinal dystrophy with progressive degeneration of the RPE, choroid, and photoreceptors (Fig. 6.7). Choroideremia is associated with the mutation of *CHM* gene encoding Rab escort protein-1 (REP-1). REP-1 protein facilitates posttranslational modification of Rab proteins regulating intracellular vesicular trafficking [26–28]. Deletion of the entire gene, small deletions, missense mutations, nonsense mutations, frameshift mutations, and splice site mutations have been identified in CHM gene [29–32], causing loss of functional domain, truncation, or absence of REP-1 [33].

The age of onset of choroideremia can be quite variable. While most patients tend to be symptomatic by the first or second decade of life and can exhibit marked progression of the nummular RPE loss across the retina from year to year even within the first decade of life [32], it is not infrequent to encounter patients who become symptomatic only in the third or fourth decade of life. Affected male patients develop night blindness with progressive patchy constriction of the peripheral visual fields and, in later disease stages, central vision loss [34, 35]. Delayed dark adaptation is anecdotally reported to be a near universal complaint of choroideremia patients, and it has been reported to be linked to reduced intraretinal rhodopsin levels [36]. However, formal large-scale investigations of night vision performance in choroideremia are lacking.

Female carriers of choroideremia usually exhibit minimal to no symptoms of the carrier state. Reports of night vision difficulties in carriers have been published, but it is not readily apparent that these reflect impaired rod function, but rather they may reflect abnormalities in rod–cone interactions in the dark-adapted state [32]. This lack of symptoms is in striking contrast with the very noticeable fundus changes female carriers of choroideremia normally exhibit, most typically a diffusely mottled and streaky pattern of coarse hyperpigmentation upon fundus examination [37, 38] (Fig. 6.8).

Similar to RP, in male patients with choroideremia, the reduction of ffERG amplitudes varies from slight reduction to undetectable [32, 39, 40]. In all patients, rod-driven ffERG responses decrease markedly, while cone responses decrease only moderately and usually follow in severity rod disease [31, 32], and the Arden ratio

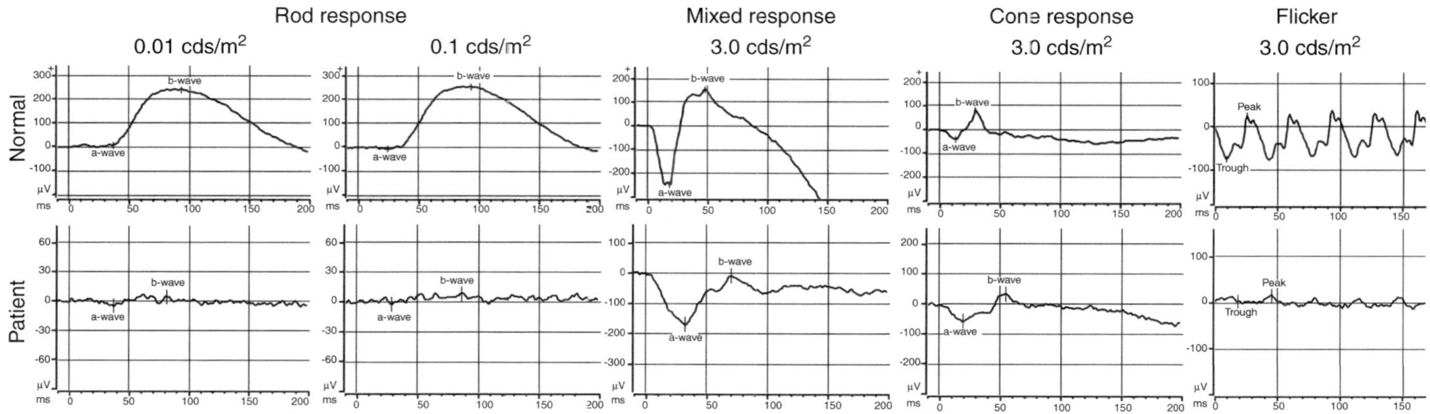

Fig. 6.6 The ffERG of ESCS. The ffERG of the patient from Fig. 6.5 exhibiting non-recordable rod responses, markedly electronegative and delayed mixed and cone responses, and markedly reduced flicker responses. The remarkable similarity in the waveform shape of the mixed and transient cone ffERG responses is typical for ESCS. In even more typical cases, the ffERG will exhibit supernormal responses with the same morphology, and this aspect too will be preserved under light-adapted conditions

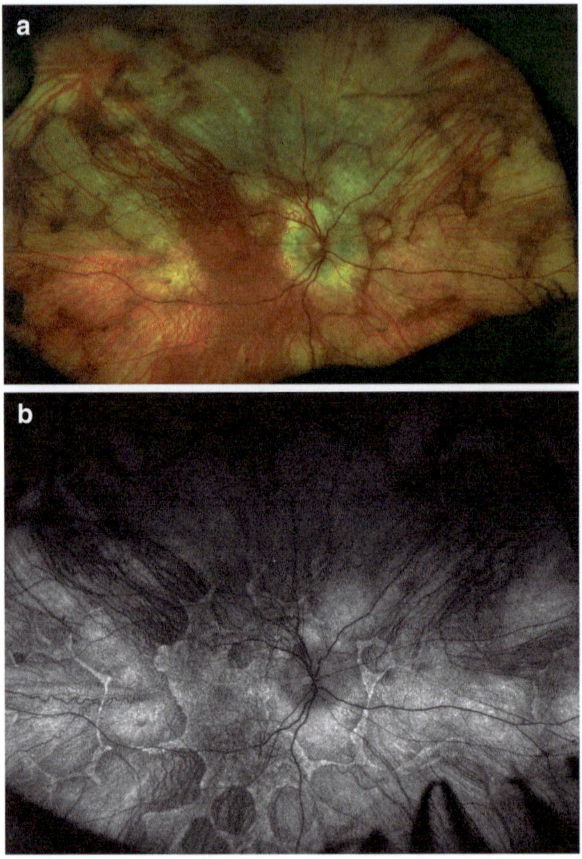

Fig. 6.7 Clinical and imaging findings in choroideremia. This 19-year-old Caucasian male with a lifelong complaint of poor night vision and negative family history. (**a**) Fundus photography shows disseminated, scalloped areas of RPE and choriocapillaris atrophy around the macula, around the disc, and in the periphery, with patchy deep RPE migration and clumping and only mildly attenuated vessels. The macular region is preserved. (**b**) Marked hypo-autofluorescence throughout the FAF scan corresponds to nummular areas of RPE and choriocapillaris atrophy with scalloped borders. There is residual near-normal autofluorescence in the macular region and disseminated small regions of intact RPE in between the areas of atrophy beyond the arcade. Molecular genetic testing corroborated the impression of choroideremia identifying a novel missense hemizygous mutation in the *CHM* gene also found in the mother, who exhibited the typical ophthalmoscopic signs of the carrier status

of the EOG is usually significantly decreased [41], reflecting the severe impairment of the health of RPE in choroideremia. However, despite the apparently pathological appearance of the fundus (Fig. 6.7), male choroideremia patients can retain robust retinal responses for a long while [32], and even in fairly widespread cases such as the example shown in Fig. 6.7, they retain readily measurable ffERG responses (Fig. 6.9).

In female carriers of choroideremia, despite the readily apparent fundus changes, the ffERG is usually normal (Fig. 6.10) or at the most mildly reduced, and the Arden

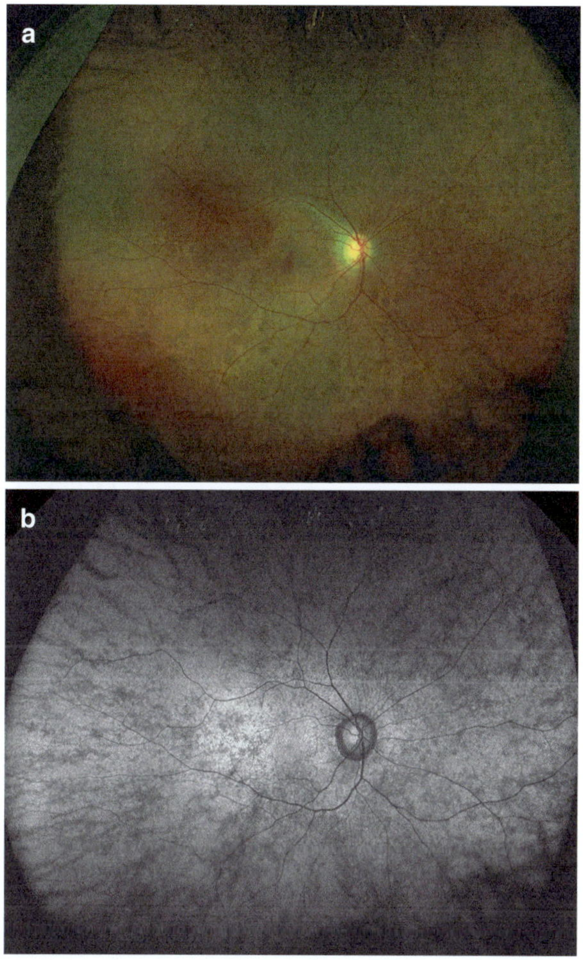

Fig. 6.8 Clinical and imaging findings in a choroideremia carrier. (**a**) Fundus photography of an Indian teenage girl reveals deep, coarse pigmentary deposits and mottling throughout the fundus. (**b**) The pigmentary deposits and mottling correspond to disseminated patches and streaks of hypo-autofluorescence, with fine hyper-autofluorescent specks at the posterior pole. Molecular genetic testing identified a multi-exon deletion in the *CHM* gene

ratio of the EOG is not significantly different from normal controls [32, 42–45]. The mfERG may exhibit amplitude reduction and/or latency prolongation in patchy fashion in female carriers of choroideremia, consistent with the patchy nature of their retinal involvement [46–48].

Gyrate Atrophy of Choroid and Retina

Gyrate atrophy of choroid and retina is an extremely rare autosomal recessive chorioretinal dystrophy caused by mutations in the *OAT1* gene encoding ornithine aminotransferase, an enzyme expressed in both the retina and the RPE [49–53]. In addition to fundus examination revealing the characteristic features of gyrate atrophy, this disorder is diagnosed primarily through detection of elevated plasma levels of ornithine

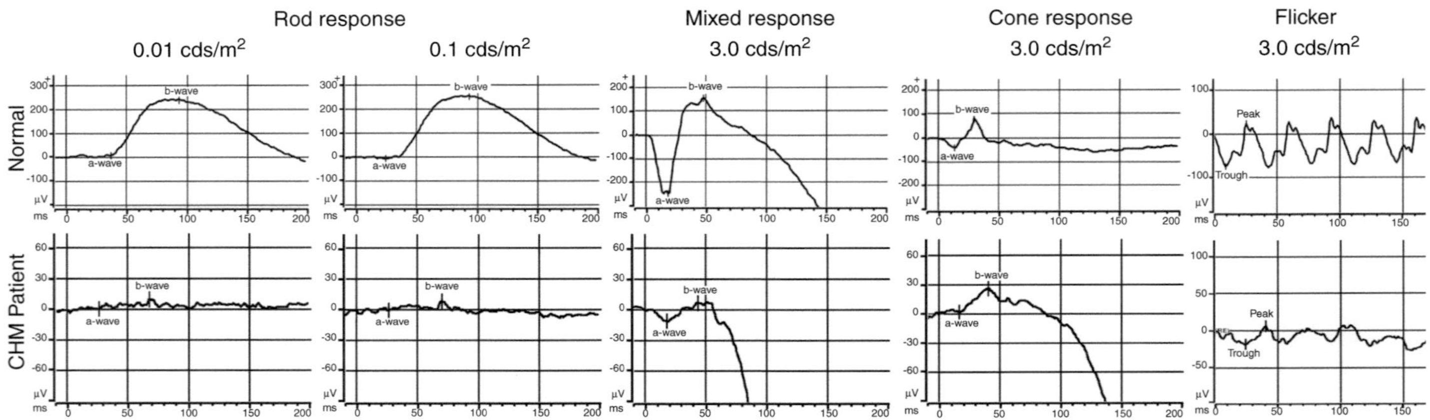

Fig. 6.9 Electroretinographic findings in choroideremia. The ffERG of a 23-year-old Caucasian male patient with choroideremia shows barely recordable rod-driven responses and markedly reduced but clearly recordable mixed and cone-driven responses. These findings of a rod–cone pattern of retinal dystrophy are not per se distinguishable from RP or other conditions with the same pattern, whereby molecular confirmation of the suspected diagnosis of choroideremia should always be sought

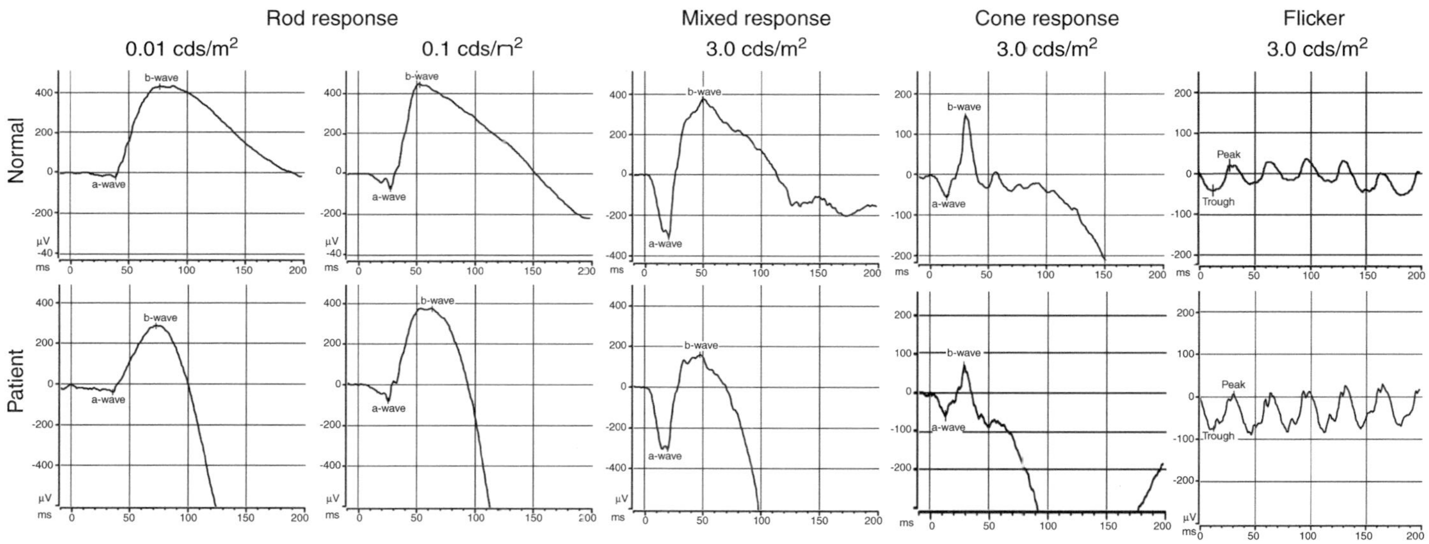

Fig. 6.10 Electroretinographic findings in a female carrier of choroideremia. The ffERG of the female carrier from Fig. 6.8 shows normal rod, mixed and cone amplitudes and timing, providing no evidence for widespread retinopathy associated with her retinal findings, as it is typical for a female choroideremia carrier

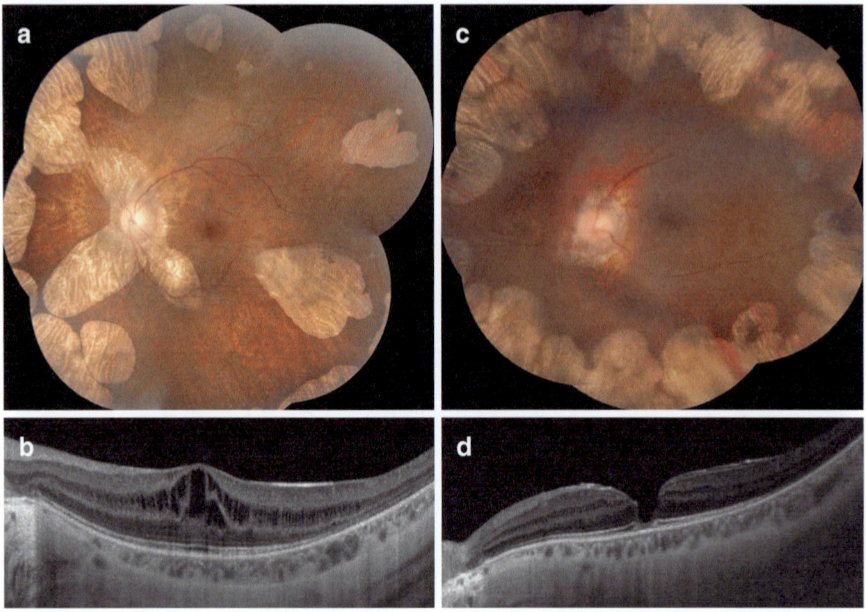

Fig. 6.11 Clinical and imaging findings in gyrate atrophy. (**a**) Fundus photo of a Patient 1 and Patient 2 (**c**) with gyrate atrophy showing coalescent well-circumscribed chorioretinal atrophy in the periphery. (**b**) OCT of patient 1 showing foveoschisis and (**d**) of patient 2 showing a lamellar hole (Adapted from Figure 1 and 3 [62])

[54–56] accumulating as the result of the enzymatic defect in ornithine aminotransferase. The development of this disease can be slowed down by long-term substantial reduction of plasma ornithine levels with arginine-restricted diet [57].

Gyrate atrophy manifests symptoms and signs that can pose a significant differential diagnostic challenge with RP and choroideremia, as they include night blindness, abnormal dark adaptation, and peripheral visual field loss [58–61]. Fundus examination, however, usually reveals a characteristic pattern of scalloped peripheral chorioretinal atrophy that, in time, progresses from the periphery toward the macula (Fig. 6.11). Associated foveoschisis and lamellar macular holes have also been reported with gyrate atrophy (Fig. 6.11) [62].

The ffERG in gyrate atrophy patients is subnormal in early stage and has undetectable rod response and low amplitudes of cone responses in late stage [60, 61, 63–65], not distinguishable from the ffERG of RP or choroideremia patients.

Late-Onset Retinal Degeneration (L-ORD)

Late-onset retinal degeneration (L-ORD) is a rare autosomal dominant late-onset rod–cone dystrophy with many clinical and pathological similarities to age-related macular degeneration (AMD) due to mutations in the *CTRP5,* a short-chain collagen gene that, when defective, results in the formation of extracellular deposits [66].

The onset of L-ORD is usually in the fifth to sixth decade with night blindness and punctate yellow-white deposits in the retina, choroidal neovascularization, and chorioretinal atrophy [67–69]. The disorder is characterized by a thick extracellular sub-RPE deposits between the basal lamina of RPE and Bruch's membrane, spreading from macula to the ora serrata [67, 68]. With disease progression, widespread loss of RPE and photoreceptors with choroidal neovascularization and disciform macular scarring occurs [67, 68, 70]. L-ORD differs from AMD in the more extensive distribution of sub-RPE deposits and atrophy, causing loss of both central and peripheral vision.

In the early stage of L-ORD, the ffERG can be normal. In later stages of L-ORD, the rod ffERG response is usually significantly decreased after a standard (30 min) period of dark adaptation, but it can improve significantly after extended dark adaptation. This is due to the characteristic delay in dark adaptation typical of L-ORD patients, even at the earliest and presymptomatic stages of the disease [69]. ffERG cone responses are mildly decreased in L-ORD. Pattern ERG is usually undetectable. However, the rod response can be improved to subnormal or normal with overnight dark adaptation [71–73].

Bietti's Crystalline Dystrophy

Bietti's crystalline dystrophy (BCD) is a rare progressive autosomal recessive disease associated with mutations in the *CYP4V2* gene, encoding for a fatty acid ω-hydroxylase of both unsaturated and saturated fatty acids that is part of the cytochrome P450 superfamily [74, 75]. This enzyme is mainly localized in the endoplasmic reticulum of RPE and plays a role in the operating of physiological lipid recycling system between the RPE and photoreceptors important for the maintenance of normal visual function [76–78].

The phenotype of BCD is extremely variable: Different mutations of the *CYP4V2* gene show different severities in the manifested phonotypes [79]. Clinically, the hallmark of BCD is the deposition of tiny crystals throughout the posterior pole of the retina (Fig. 6.12) and at the corneal limbus, followed by the degeneration of RPE, choriocapillaris, and photoreceptors that may also cause choroidal neovascularization (as in the right eye of the example shown in Fig. 6.12), sclerosis of the choroidal vessels, and atrophy of RPE. In time, the characteristic crystals tend to fade and can become no longer appreciable either at the limbus or at the retinal levels, in part also due to the progressive atrophy of the RPE these patients can develop, tending to be patchy and later on coalesce [80].

BCD patients often start complaining of night blindness, reduced visual acuity, and abnormal color vision between the second to third decade of life, although both earlier and later onset have been reported [81]. Paracentral scotoma or peripheral visual field defect and legal blindness can occur in the fifth to sixth decade of life of BCD patients.

Electroretinographic tests can show the progressive dysfunction of rod and cone photoreceptors in patients with BCD. The amplitudes of all ffERG responses can

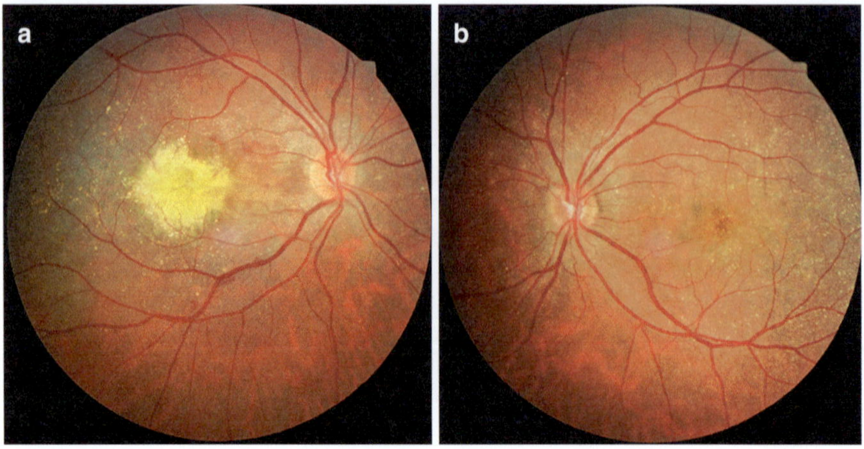

Fig. 6.12 Fundus findings in Bietti's crystalline dystrophy. (**a**, **b**) Right and left eye, respectively, of an Italian female patient in her 30s with molecularly confirmed Bietti's crystalline dystrophy (compound heterozygote mutations in the *CYP4V2* gene) showing disseminated tiny crystalline deposits in retina. The right eye developed a choroidal neovascular membrane, and this photo was taken a month after administration of an intravitreal injection of bevacizumab ultimately successful in resolving the complication and leading to significant visual recovery

range from near normal initially to undetectable in the late disease stages. In mfERG, decreased waveform densities and increased latencies are found in the entire tested field. EOG Arden radio ranges from normal to significantly decreased. The pattern ERG ranges from normal to slightly reduced in amplitude and delayed depending on the amount of macular disease severity [79, 80, 82–97].

Cone Dystrophies

Cone dystrophies are a group of rare, genetically heterogeneous retinal diseases characterized by dysfunction and degeneration of cone photoreceptors. The general symptoms include decreased visual acuity, photophobia, central scotomas, and abnormal color vision. Cone dystrophies can be inherited in an autosomal recessive, autosomal dominant, or X-linked fashion [98–100]. This group of conditions has been broadly classified into two groups: stationary and progressive. The stationary forms have been reviewed alongside other stationary or minimally progressive disorders under section 4.1. In progressive cone dystrophies, the symptoms gradually worsen over time, while of the onset of the symptoms varies from childhood to late adulthood. Eventually, rod dysfunction may also develop, broadening the dysfunction to a more widespread cone–rod dystrophy (CORD) phenotype [98–100]. Some syndromes, and in particular spinocerebellar ataxia type 7 (SCA7), can also be associated with cone or cone–rod dystrophy [101].

The ffERG of the cone dystrophy patient reveals moderate to severe reduction of light-adapted ERG responses, while the dark-adapted ERG responses are normal or only slightly decreased (Figs. 6.13 and 6.14). The mfERG usually shows severe reduction of the response density across the board (Figs. 6.15 and 6.16). In more focal forms, however, the peripheral responses can be relatively better preserved. Visual fields usually show central scotomas, and autofluorescence and fluorescein angiography may or may not show central changes.

Cone–Rod Dystrophies

Cone–rod dystrophies (CORDs, prevalence 1:40,000), by definition, result from a loss of cone cells, concomitantly to, or followed by loss of rods, and this particular diagnosis is usually established on the basis of the ratio of rod and cone compromise to ffERG testing or to other types of photoreceptor-specific psychophysical methods. There is significant genetic heterogeneity among CORDs [102]. One of the most common forms of CORD is inherited in autosomal recessive fashion and caused by mutations in the *ABCA4* gene, the same responsible for Stargardt's disease, representing a mere continuum into more severe forms of "Stargardt plus" disease [103, 104]. Over 20 genes responsible for CORDs have been identified [2], including several dominant ones (Fig. 6.17) and an X-linked form linked to mutations in the ORF15 region of the *RPGR* gene [105–107] (Fig. 6.18) that is otherwise typically responsible for the majority of the X-linked RP cases.

The age of the onset of the CORDs varies from childhood to late adulthood [108]. Affected patients experience loss of visual acuity, central scotomas, photophobia, and impaired color vision, associated with night blindness and peri- and para-central visual field scotomas too. In some extreme cases, the patients have retinal pigmentation and chorioretinal atrophy of both the central and peripheral retina. The signs and symptoms of cone–rod dystrophy usually start from the central area of the retina with cone>rod functional compromise to ffERG testing since the early stages of the disease. In the late stage of CORDs, both cone and rod cells have severe degeneration, and the ffERG responses are all severely decreased (Fig. 6.19). The mfERG amplitudes in CORD are consistently markedly decreased, and they become more so with advancing disease (Fig. 6.20). Patients with varying degrees of severity may have different phenotypic signs (Figs. 6.21, 6.22, and 6.23). CORDs are typically non-syndromic, but they may also be caused by some syndromes, such as the aforementioned SCA7 [101] and the Alström syndrome [109, 110] (see below).

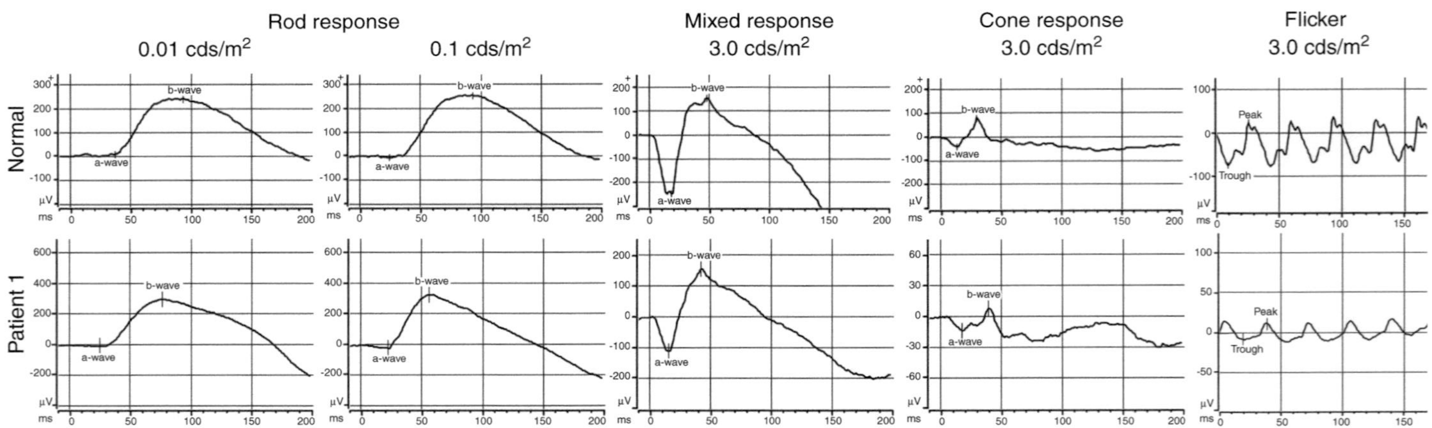

Fig. 6.13 Electroretinographic findings in autosomal recessive cone dystrophy. A 35-year-old Indian woman with autosomal recessive cone dystrophy caused by compound heterozygosity for a missense and a frameshift mutation in the *POC1B* gene. The ffERG shows normal rod responses, slightly reduced a-wave to mixed stimuli, and reduced and delayed cone ERGs responses indicative of a cone-selective retinopathy

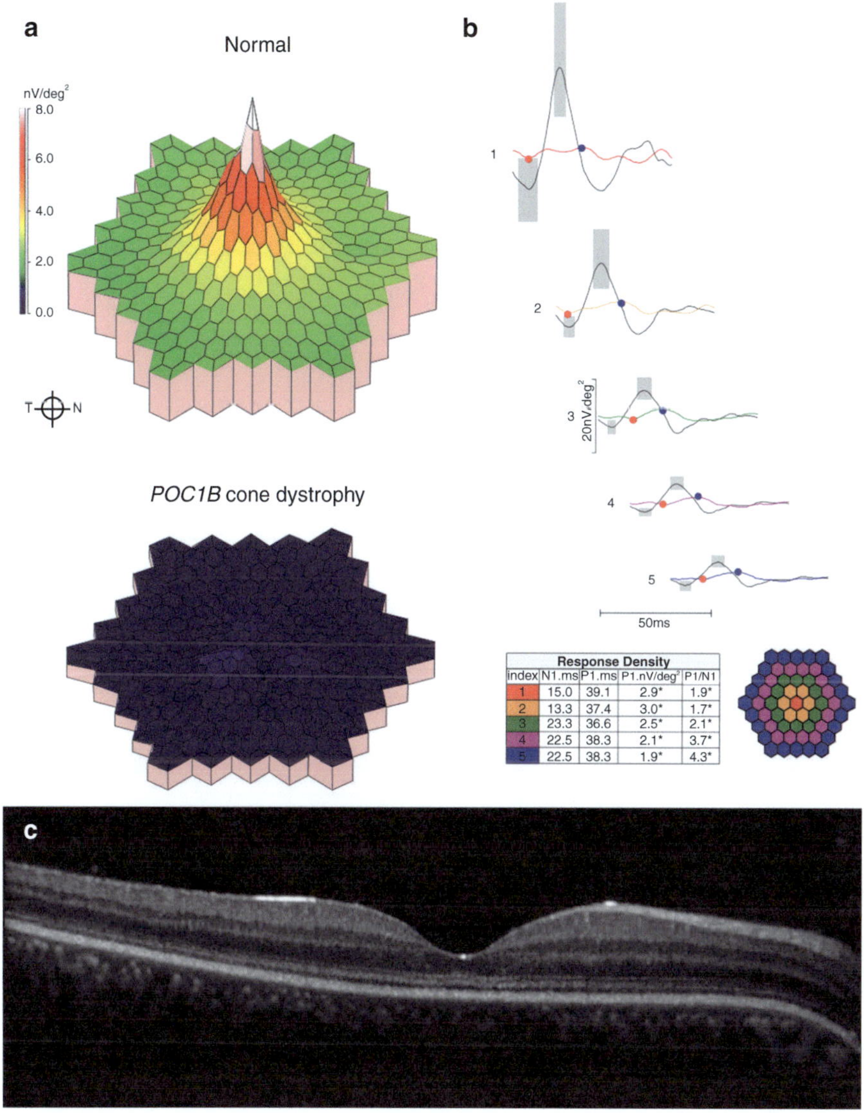

Response Density				
index	N1.ms	P1.ms	P1.nV/deg²	P1/N1
1	15.0	39.1	2.9*	1.9*
2	13.3	37.4	3.0*	1.7*
3	23.3	36.6	2.5*	2.1*
4	22.5	38.3	2.1*	3.7*
5	22.5	38.3	1.9*	4.3*

Fig. 6.14 Macular functional and imaging findings in autosomal recessive cone dystrophy. (**a, b**) Multifocal ERG density plot and concentric ring trace arrays (the grey bars indicate the normal ranges of N1 and P1 components with 95% confidence intervals, respectively) of the patient in Fig. 6.13 (POC1B mutations) showing markedly reduced and delayed responses, indicating severe macular cone compromise, in spite of (**c**) a macular OCT mainly remarkable only for a grainy EZ and an accentuated interdigitation zone, seen as a fine hyporeflective cleft all across the macular scan

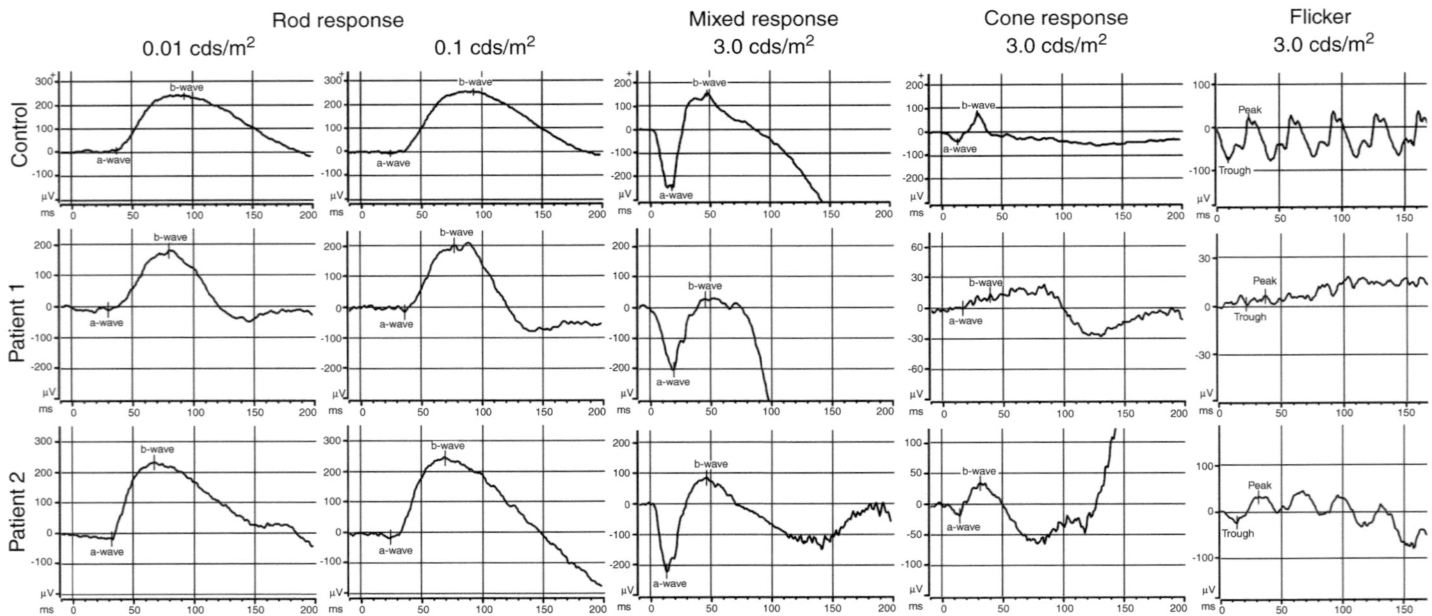

Fig. 6.15 Electroretinographic findings in dominant cone dystrophy, compared with normal values. ffERG findings on two African-American patients, PT1 (31-year-old female) and PT2 (her daughter, 11 years old), both symptomatic for photophobia and reduced acuity and diagnosed with autosomal dominant cone dystrophy with a mutation in the GUCY2D gene. While both have normal rod responses, PT1 has severely reduced cone responses, while her daughter (PT2) has borderline low and mildly delayed cone responses, consistent with an earlier disease stage at her younger age

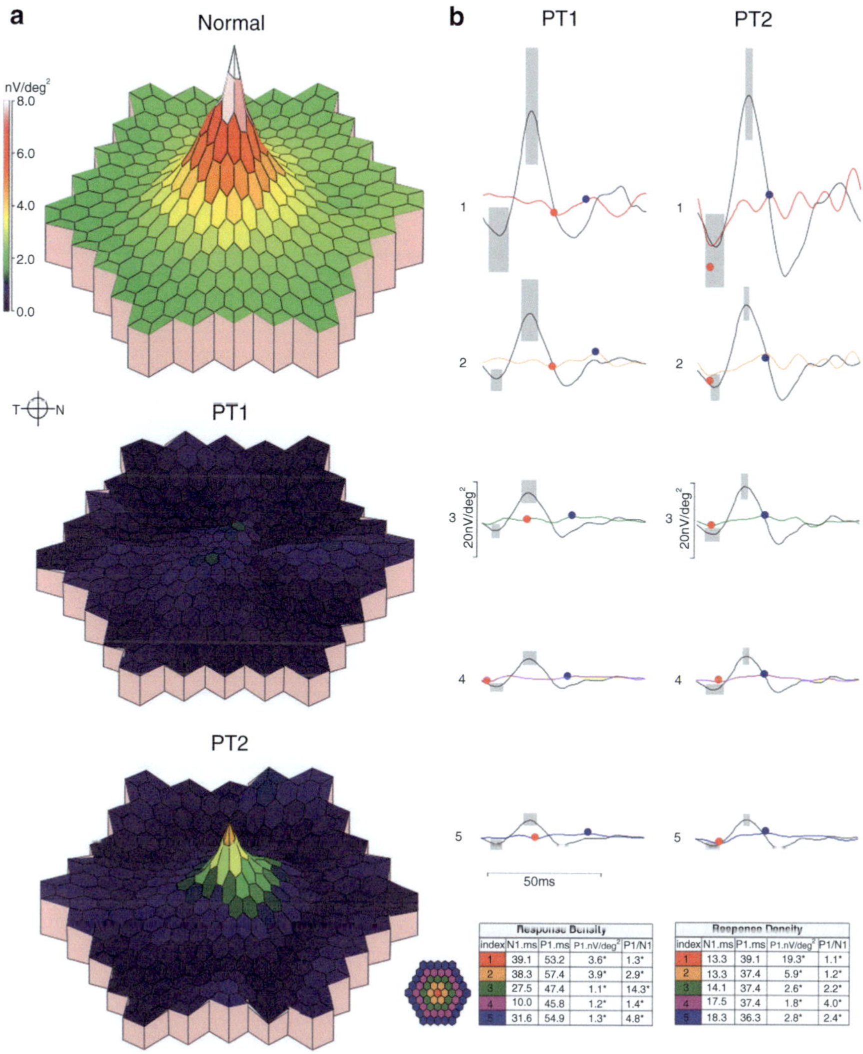

Response Density				
index	N1.ms	P1.ms	P1.nV/deg²	P1/N1
1	39.1	53.2	3.6*	1.3*
2	38.3	57.4	3.9*	2.9*
3	27.5	47.4	1.1*	14.3*
4	10.0	45.8	1.2*	1.4*
5	31.6	54.9	1.3*	4.8*

Response Density				
index	N1.ms	P1.ms	P1.nV/deg²	P1/N1
1	13.3	39.1	19.3*	1.1*
2	13.3	37.4	5.9*	1.2*
3	14.1	37.4	2.6*	2.2*
4	17.5	37.4	1.8*	4.0*
5	18.3	36.3	2.8*	2.4*

Fig. 6.16 Macular function in dominant cone dystrophy, compared with normal values. Comparison of the mfERG response density plots (**a**) and eccentricity ring average responses (**b**) for the cases (PT1 and PT2) shown in Fig. 6.14. PT1 has markedly reduced and delayed responses, whereas the daughter (PT2) has a discernable foveal peak, albeit reduced compared to normal, with overall detectable and less delayed responses (**a** density plot, **b** concentric ring trace arrays). This difference reflects well their best corrected visual acuities of 20/200 in PT1 and 20/32 in PT2

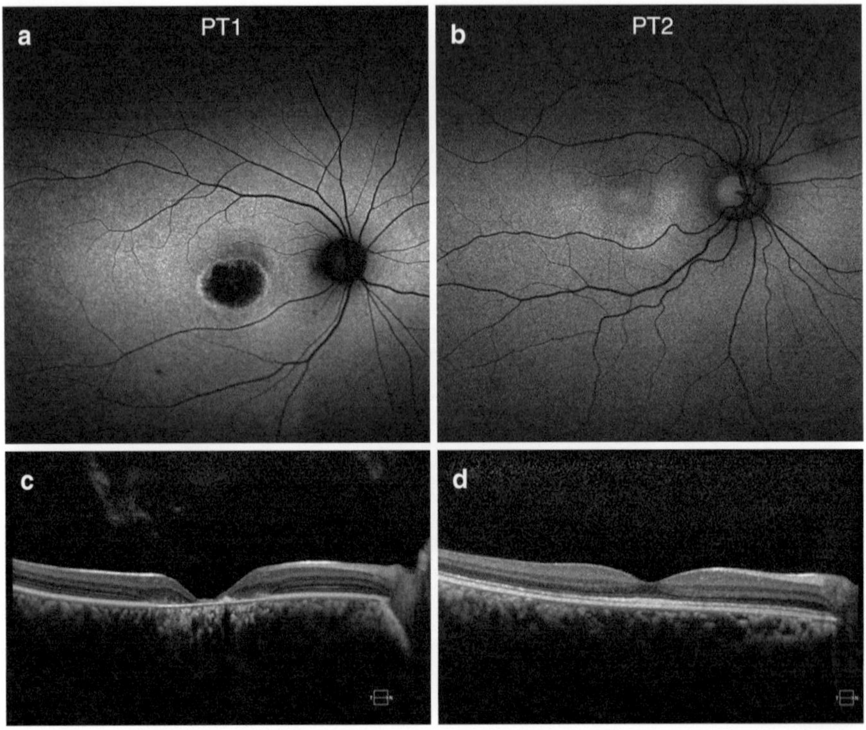

Fig. 6.17 Imaging findings in dominant cone rod dystrophy. (**a**, **b**) Fundus autofluorescence of PT1 and PT2, respectively. (**c**, **d**) OCT findings in the same two patients. PT1 has definite discrete central hypo-autofluorescence surrounded by a hyper-autofluorescent rim, the inner edge of which corresponds to the area of marked macular retinal thinning and RPE loss (hyper-transmission defects) seen on OCT. PT2 only has faint central FAF mottling and trace hypo- autofluorescence around the fovea. Her OCT shows increased signal intensity at the foveal EZ level and mild diffuse EZ fragmentation

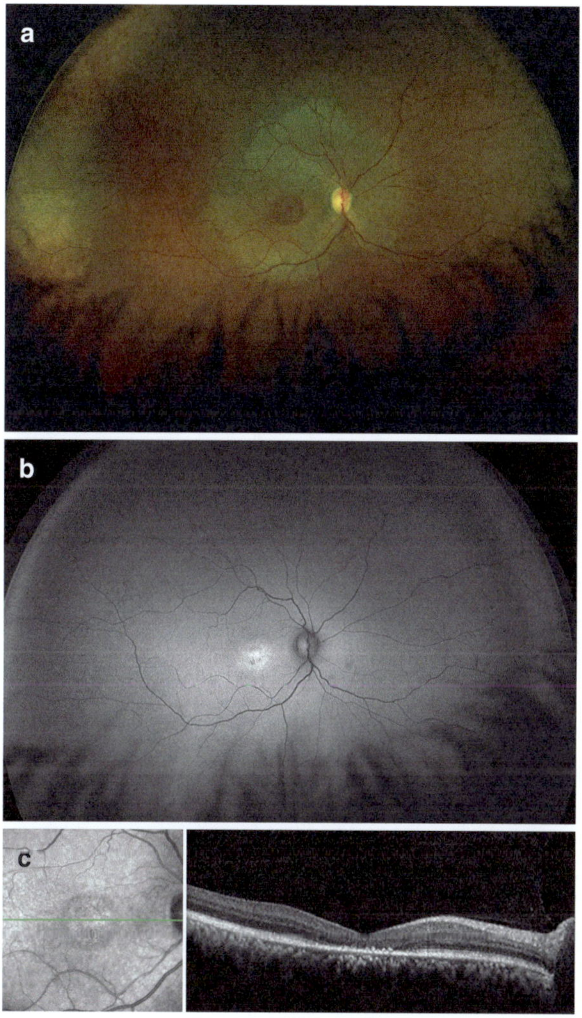

Fig. 6.18 Clinical and imaging findings in X-linked cone rod dystrophy with tapetal-like sheen. (**a**) A fundus photo of a 44-year-old African-American man with a recent complaint of declining acuity, difficulties reading, and photophobia shows a metallic sheen (tapetal-like reflex or TLR) area across the entire posterior pole and focal RPE changes in the parafoveal area. Additional areas of anomalous retinal reflexes can be seen also in the far temporal periphery. (**b**) Fundus autofluorescence shows faint hyper-autofluorescence throughout the posterior pole, focally increased autofluorescence foveally and perifoveally, with small hypo-autofluorescent spots near the fovea. (**c**) The OCT shows retinal thinning, severe ELM loss, and EZ loss and crumbling at the fovea, with hyper-transmission defects resulting from RPE dropout. The diagnosis of X-linked cone dystrophy with tapetal-like sheen was confirmed molecularly by identifying a nonsense mutation in the ORF15 portion of the *RPGR* gene

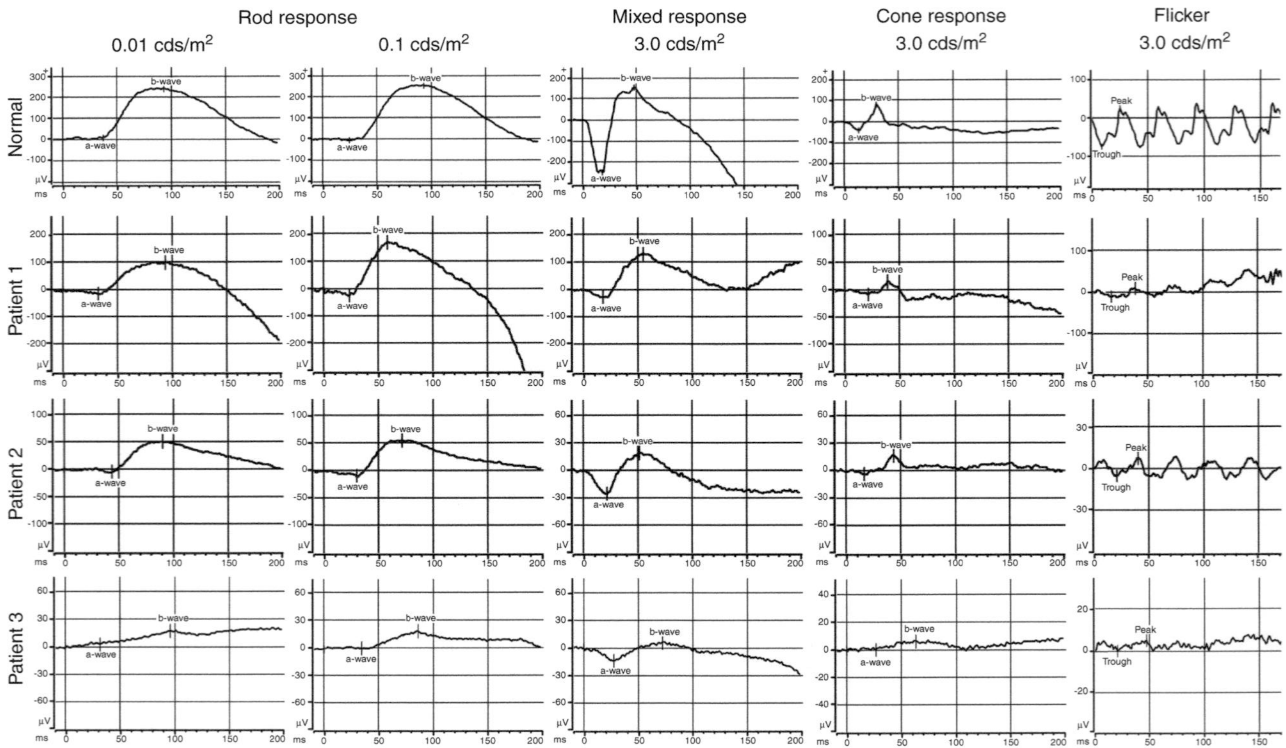

Fig. 6.19 Electroretinographic findings in patients with cone–rod dystrophy. ERG of three patients with cone–rod dystrophy (CORD) at different stages of disease. Patient 1 has normal mildly reduced rod responses, markedly decreased mixed a-wave, and moderately reduced cone responses. Patient 2 has moderately reduced rod and mixed responses and severely reduced but detectable cone ERGs OU. Patient 3 has markedly reduced but clearly recordable dark-adapted responses and barely recordable cone response, consistent with the diagnosis of advanced CORD

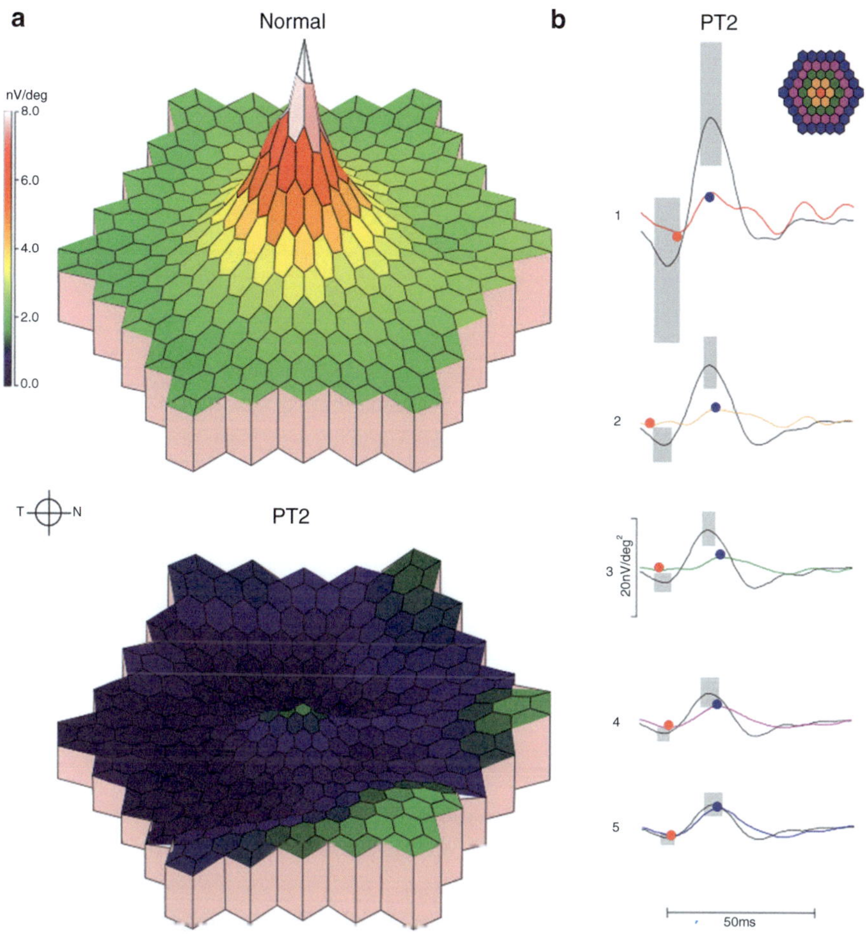

Fig. 6.20 Macular function in cone–rod dystrophy. The mfERG of Patient 2 from Fig. 6.19 (a density plot, b concentric ring trace arrays) shows markedly reduced responses in pericentral pattern with near normal peripheral ring responses and detectable but markedly attenuated foveal peak response density

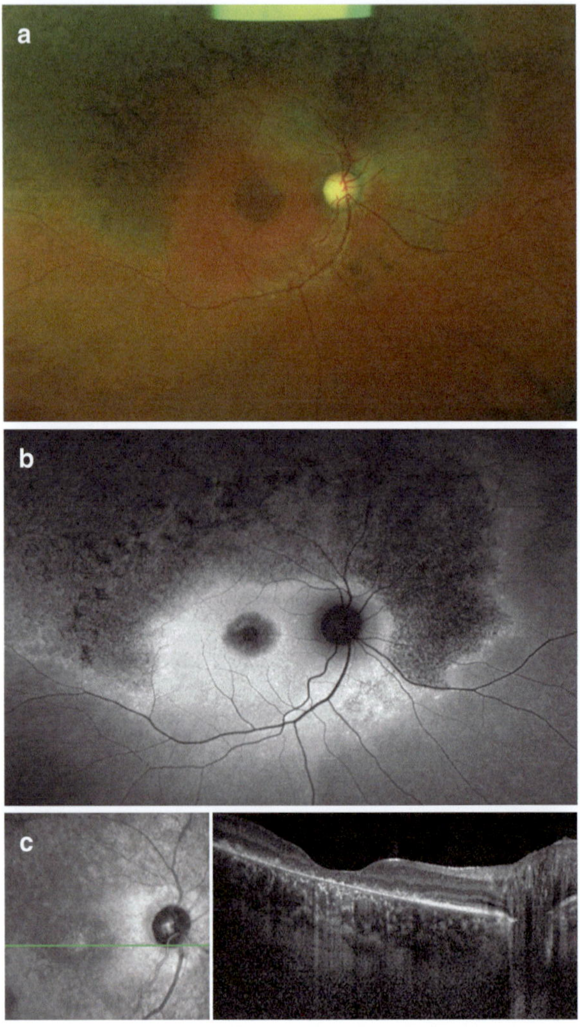

Fig. 6.21 Clinical and imaging findings in atypical *ABCA4*-linked cone–rod dystrophy. (**a**) Fundus photo of a 10-year-old African-American boy with compound heterozygous *ABCA4* mutations (a nonsense and a deep intronic one), causing a severe, early-onset form of atypical cone–rod dystrophy (CORD) with features of both Stargardt's disease and retinitis pigmentosa. Central RPE changes, loss of foveal reflex, and bone spicule-like pigment deposits at the superior arcade are seen. (**b**) On fundus autofluorescence, there is macular hypo-autofluorescence and a hemi-ring of superior hypo-autofluorescence along the superotemporal arcade. There is also diffuse hyper-autofluorescence across the posterior pole and punctate/nummular hyper-autofluorescence all along the edges of the areas of hypo-autofluorescence. (**c**) On OCT, there is marked foveal retinal thinning, ellipsoid zone disruption and crumbling, and marked hyper-transmission defects due to macular RPE atrophy

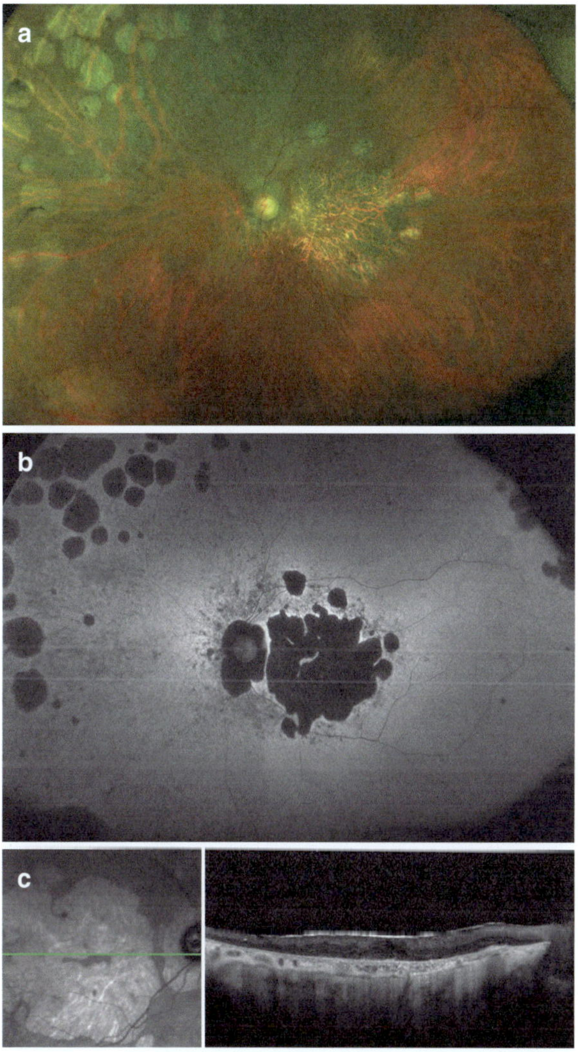

Fig. 6.22 Clinical and imaging findings in *CDHR1*-associated cone–rod dystrophy. (**a**) Fundus photo of a 65-year-old Caucasian male with a slowly progressive pericentral retinal dystrophy with a cone–rod dystrophy (CORD) pattern by ffERG criteria, caused by compound heterozygote mutations in the *CDHR1* gene (a synonymous disease-causing change and a 7-base pair deletion, both previously reported). Central RPE atrophy, mild attenuated vessels, and disseminated far peripheral patches of atrophic RPE are seen. (**b**) On fundus autofluorescence, there are coalescent central and disseminated far peripheral hypo-autofluorescent spots corresponding to the patches of the atrophic RPE, with a ring of speckled hyper-AF around the macular lesions. Note the lack of peripapillary sparing, the hallmark of most cases of *ABCA4*-linked Stargardt's disease, and more aggressive phenotypes such as the one shown in Fig. 6.21. The small linear autofluorescent residues in the macular region accounts for a sliver of central vision this patient retains. (**c**) An OCT scan shows marked EZ loss throughout with a small, thin foveal residue thereof, retinal thinning perifoveally, and lack of a foveal pit due to tangential traction from an epiretinal membrane exerting a tenting effect

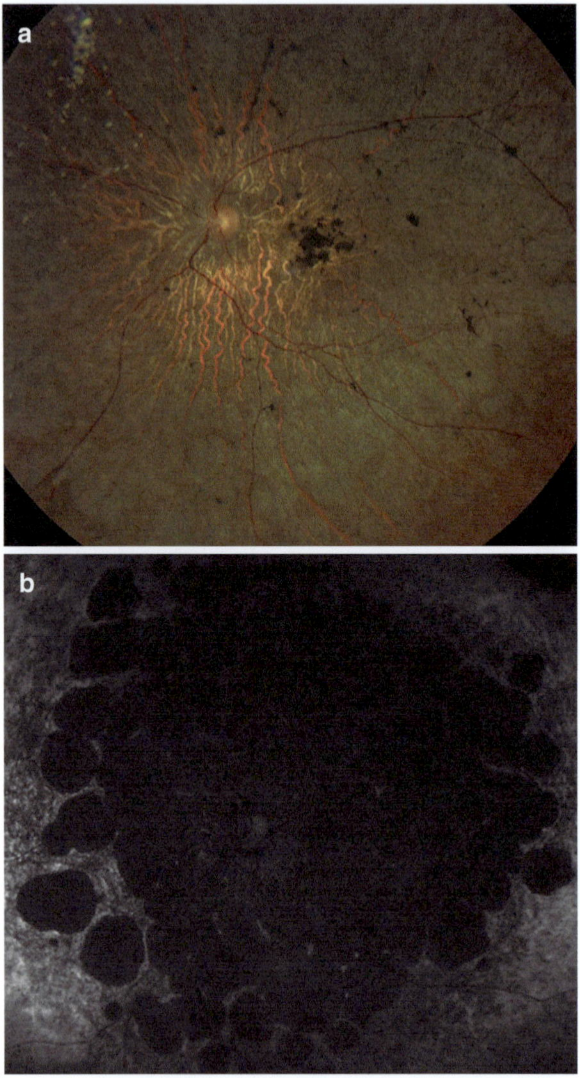

Fig. 6.23 Clinical and imaging findings in advanced ABCA4-linked cone–rod dystrophy. (**a**) Fundus photo of a 66-year-old Caucasian woman affected with advanced cone–rod dystrophy (CORD) due to compound heterozygous *ABCA4* mutations, showing a large area of RPE atrophy, macular pigmentary deposits and retinal thinning throughout the posterior pole in a confluent nummular pattern, severe peripapillary atrophy, and markedly attenuated vessels. Some vitreal deposits due to asteroid hyalosis can also be seen in the superonasal quadrant. (**b**) Fundus autofluorescence of the same patient showing severe hypo-autofluorescence throughout the posterior pole and beyond the arcades, with a ring of peripheral preservation on autofluorescence. The ffERG of this patient (not shown) showed severe reduction of cone > rod responses and fits the criteria for a type III patient of the ABCA4-related Stargardt disease classification, emphasizing the continuum existing between these conditions

References

1. Iannaccone A, Berdia J. Retinitis pigmentosa. 2017. Review No. 21. Danbury, CT:National Organization for Rare Disorders, Inc.; www.rarediseases.org
2. RetNet – Retinal Information Network., https://sph.uth.edu/retnet/home.htm. 2019.
3. Nagy D, et al. Long-term follow-up of retinitis pigmentosa patients with multifocal electroretinography. Invest Ophthalmol Vis Sci. 2008;49(10):4664–71.
4. Walia S, et al. Visual acuity in patients with Leber's congenital amaurosis and early childhood-onset retinitis pigmentosa. Ophthalmology. 2010;117(6):1190–8.
5. den Hollander AI, et al. Leber congenital amaurosis: genes, proteins and disease mechanisms. Prog Retin Eye Res. 2008;27(4):391–419.
6. Stone EM. Leber congenital amaurosis – a model for efficient genetic testing of heterogeneous disorders: LXIV Edward Jackson Memorial Lecture. Am J Ophthalmol. 2007;144(6):791–811.
7. Pennesi ME, et al. Residual electroretinograms in young Leber congenital amaurosis patients with mutations of AIPL1. Invest Ophthalmol Vis Sci. 2011;52(11):8166 73.
8. Occelli LM, et al. CrxRdy cat: a large animal model for CRX-associated Leber congenital Amaurosis. Invest Ophthalmol Vis Sci. 2016;57(8):3780–92.
9. Chang B. Mouse models as tools to identify genetic pathways for retinal degeneration, as exemplified by Leber's congenital Amaurosis. Methods Mol Biol. 2016;1438:417–30.
10. Khan AO, et al. Peripherin mutations cause a distinct form of recessive Leber congenital amaurosis and dominant phenotypes in asymptomatic parents heterozygous for the mutation. Br J Ophthalmol. 2016;100(2):209–15.
11. Kuniyoshi K, et al. Longitudinal clinical course of three Japanese patients with Leber congenital amaurosis/early-onset retinal dystrophy with RDH12 mutation. Doc Ophthalmol. 2014;128(3):219–28.
12. Iannaccone A, et al. Treatment of adult-onset acute macular retinoschisis in enhanced S-cone syndrome with oral acetazolamide. Am J Ophthalmol. 2009;147:307–12.
13. Sharon D, et al. Shared mutations in NR2E3 in enhanced S-cone syndrome, Goldmann-Favre syndrome, and many cases of clumped pigmentary retinal degeneration. Arch Ophthalmol. 2003;121(9):1316–23.
14. Milam AH, et al. The nuclear receptor NR2E3 plays a role in human retinal photoreceptor differentiation and degeneration. Proc Natl Acad Sci U S A. 2002;99(1):473–8.
15. Haider NB, et al. Mutation of a nuclear receptor gene, NR2E3, causes enhanced S cone syndrome, a disorder of retinal cell fate. Nat Genet. 2000;24(2):127–31.
16. Hood DC, et al. Enhanced S cone syndrome: evidence for an abnormally large number of S cones. Vision Res. 1995;35(10):1473–81.
17. Haider NB, Naggert JK, Nishina PM. Excess cone cell proliferation due to lack of a functional NR2E3 causes retinal dysplasia and degeneration in rd7/rd7 mice. Hum Mol Genet. 2001;10(16):1619–26.
18. Pachydaki SI, et al. Long-term follow-up in enhanced s-cone syndrome. Retin Cases Brief Rep. 2009;3(2):118–20.
19. Genead MA, Fishman GA, McAnany JJ. Efficacy of topical dorzolamide for treatment of cystic macular lesions in a patient with enhanced S-cone syndrome. Doc Ophthalmol. 2010;121(3):231–40.
20. Kiszkielis M, Lubinski W, Penkala K. Topical dorzolamide treatment of macular cysts in the enhanced S-cone syndrome patient. Doc Ophthalmol. 2013;126(3):241–6.
21. Marmor MF, et al. Diagnostic clinical findings of a new syndrome with night blindness, maculopathy, and enhanced S cone sensitivity. Am J Ophthalmol. 1990;110(2):124–34.
22. Jacobson SG, et al. SWS (blue) cone hypersensitivity in a newly identified retinal degeneration. Invest Ophthalmol Vis Sci. 1990;31(5):827–38.
23. Kuniyoshi K, et al. New truncation mutation of the NR2E3 gene in a Japanese patient with enhanced S-cone syndrome. Jpn J Ophthalmol. 2016;60(6):476–85.

24. Kuniyoshi K, et al. Novel mutations in enhanced S-cone syndrome. Ophthalmology. 2013;120(2):431 e1–6.
25. Audo I, et al. Phenotypic variation in enhanced S-cone syndrome. Invest Ophthalmol Vis Sci. 2008;49(5):2082–93.
26. Seabra MC, et al. Purification of component A of Rab geranylgeranyl transferase: possible identity with the choroideremia gene product. Cell. 1992;70(6):1049–57.
27. van den Hurk JA, et al. Molecular basis of choroideremia (CHM): mutations involving the Rab escort protein-1 (REP-1) gene. Hum Mutat. 1997;9(2):110–7.
28. Pfeffer SR. Rab GTPases: master regulators of membrane trafficking. Curr Opin Cell Biol. 1994;6(4):522–6.
29. Sanchez-Alcudia R, et al. A comprehensive analysis of choroideremia: from genetic characterization to clinical practice. PLoS One. 2016;11(4):e0151943.
30. Ben Charfeddine I, et al. Genetic study in a tunisian family revealed IVS1+1G>A mutation in the CHM gene. Ann Biol Clin (Paris). 2015;73(4):469–73.
31. Contestabile MT, et al. Clinical and genetic studies in a family with a new splice-site mutation in the choroideremia gene. Mol Vis. 2014;20:325–33.
32. Mura M, et al. Clinical and functional findings in choroideremia due to complete deletion of the CHM gene. Arch Ophthalmol. 2007;125(8):1107–13.
33. Sergeev YV, et al. The functional effect of pathogenic mutations in Rab escort protein 1. Mutat Res. 2009;665(1–2):44–50.
34. Roberts MF, et al. Retrospective, longitudinal, and cross sectional study of visual acuity impairment in choroideraemia. Br J Ophthalmol. 2002;86(6):658–62.
35. MacDonald IM, et al. Choroideremia. Gene Reviews 2008 May 28, 2008 April 15, 2009]. Available from: http://www.ncbi.nlm.nih.gov/bookshelf/br.fcgi?book=gene&part=choroid#choroid.grID19666.
36. Fulton AB, Hansen RM. The relation of rhodopsin and scotopic sensitivity in choroideremia. Am J Ophthalmol. 1987;104(5):524–32.
37. MacDonald IM, et al. Histopathology of the retinal pigment epithelium of a female carrier of choroideremia. Can J Ophthalmol. 1997;32(5):329–33.
38. Perez-Cano HJ, Garnica-Hayashi RE, Zenteno JC. CHM gene molecular analysis and X-chromosome inactivation pattern determination in two families with choroideremia. Am J Med Genet A. 2009;149A(10):2134–40.
39. Ponjavic V, et al. Phenotype variations within a choroideremia family lacking the entire CHM gene. Ophthalmic Genet. 1995;16(4):143–50.
40. Renner AB, et al. Choroideremia: variability of clinical and electrophysiological characteristics and first report of a negative electroretinogram. Ophthalmology. 2006;113(11):2066 e1–10.
41. Chen MS, et al. Blood-aqueous barrier function in a patient with choroideremia. J Formos Med Assoc. 2010;109(2):167–71.
42. Renner AB, et al. Progression of retinal pigment epithelial alterations during long-term follow-up in female carriers of choroideremia and report of a novel CHM mutation. Arch Ophthalmol. 2009;127(7):907–12.
43. Iino Y, et al. A novel mutation (967-970+2)delAAAGGT in the choroideremia gene found in a Japanese family and related clinical findings. Jpn J Ophthalmol. 2008;52(4):289–97.
44. Sieving PA, Niffenegger JH, Berson EL. Electroretinographic findings in selected pedigrees with choroideremia. Am J Ophthalmol. 1986;101(3):361–7.
45. Yau RJ, et al. Choroideremia carriers maintain a normal electro-oculogram (EOG). Doc Ophthalmol. 2007;114(3):147–51.
46. Vajaranant TS, et al. Detection of mosaic retinal dysfunction in choroideremia carriers electroretinographic and psychophysical testing. Ophthalmology. 2008;115(4):723–9.
47. Preising MN, et al. Fundus autofluorescence in carriers of choroideremia and correlation with electrophysiologic and psychophysical data. Ophthalmology. 2009;116(6):1201–9 e1-2.
48. Cheung MC, et al. Detection of localized retinal dysfunction in a choroideremia carrier. Am J Ophthalmol. 2004;137(1):189–91.

49. Wu J, et al. The ornithine aminotransferase (OAT) locus is linked and distal to D10S20 on the long arm of chromosome 10. Cytogenet Cell Genet. 1988;48(2):126–7.
50. Hayden MR, et al. A polymorphic DNA marker that represents a conserved expressed sequence in the region of the Huntington disease gene. Am J Hum Genet. 1988;42(1):125–31.
51. Mitchell GA, et al. Human ornithine-delta-aminotransferase. cDNA cloning and analysis of the structural gene. J Biol Chem. 1988;263(28):14288–95.
52. Rao GN, Cotlier E. Ornithine delta-aminotransferase activity in retina and other tissues. Neurochem Res. 1984;9(4):555–62.
53. Ratzlaff K, Baich A. Comparison of ornithine aminotransferase activities in the pigment epithelium and retina of vertebrates. Comp Biochem Physiol B. 1987;88(1):35–7.
54. Takki K. Gyrate atrophy of the choroid and retina associated with hyperornithinaemia. Br J Ophthalmol. 1974;58(1):3–23.
55. Takki K, Simell O. Genetic aspects in gyrate atrophy of the choroid and retina with hyperornithinaemia. Br J Ophthalmol. 1974;58(11):907–16.
56. Simell O, Takki K. Raised plasma-ornithine and gyrate atrophy of the choroid and retina. Lancet. 1973;1(7811):1031–3.
57. Kaiser-Kupfer MI, Caruso RC, Valle D. Gyrate atrophy of the choroid and retina: further experience with long-term reduction of ornithine levels in children. Arch Ophthalmol. 2002;120(2):146–53.
58. Feldman RB, et al. Epiretinal membranes and cystoid macular edema in gyrate atrophy of the choroid and retina. Retina. 1989;9(2):139–42.
59. Takki KK, Milton RC. The natural history of gyrate atrophy of the choroid and retina. Ophthalmology. 1981;88(4):292–301.
60. Renner AB, et al. Gyrate atrophy: clinical and genetic findings in a female without arginine-restricted diet during her first 39 years of life and report of a new OAT gene mutation. Doc Ophthalmol. 2012;125(1):81–9.
61. Peltola KE, et al. Ophthalmologic heterogeneity in subjects with gyrate atrophy of choroid and retina harboring the L402P mutation of ornithine aminotransferase. Ophthalmology. 2001;108(4):721–9.
62. Braham IZ, et al. Multimodal imaging of foveoschisis and macular pseudohole associated with gyrate atrophy: a family report. BMC Ophthalmol. 2018;18(1):89.
63. Katagiri S, et al. OAT mutations and clinical features in two Japanese brothers with gyrate atrophy of the choroid and retina. Doc Ophthalmol. 2014;128(2):137–48.
64. Mehta MC, et al. Gyrate atrophy of the choroid and retina in a 5-year-old girl. Acta Ophthalmol. 1991;69(6):810–4.
65. Raitta C, Carlson S, Vannas-Sulonen K. Gyrate atrophy of the choroid and retina: ERG of the neural retina and the pigment epithelium. Br J Ophthalmol. 1990;74(6):363–7.
66. Hayward C, et al. Mutation in a short-chain collagen gene, CTRP5, results in extracellular deposit formation in late-onset retinal degeneration: a genetic model for age-related macular degeneration. Hum Mol Genet. 2003;12(20):2657–67.
67. Kuntz CA, et al. Sub-retinal pigment epithelial deposits in a dominant late-onset retinal degeneration. Invest Ophthalmol Vis Sci. 1996;37(9):1772–82.
68. Milam AH, et al. Dominant late-onset retinal degeneration with regional variation of sub-retinal pigment epithelium deposits, retinal function, and photoreceptor degeneration. Ophthalmology. 2000;107(12):2256–66.
69. Jacobson SG, et al. Phenotypic marker for early disease detection in dominant late-onset retinal degeneration. Invest Ophthalmol Vis Sci. 2001;42(8):1882–90.
70. Duvall J, et al. Extensive subretinal pigment epithelial deposit in two brothers suffering from dominant retinitis pigmentosa. A histopathological study. Graefes Arch Clin Exp Ophthalmol. 1986;224(3):299–309.
71. Papastavrou VT, et al. Improvement of retinal function in L-ORD after prolonged dark adaptation. Can J Ophthalmol. 2015;50(2):112–8.
72. Soumplis V, et al. Phenotypic findings in C1QTNF5 retinopathy (late-onset retinal degeneration). Acta Ophthalmol. 2013;91(3):e191–5.

73. Vincent A, et al. The characterization of retinal phenotype in a family with C1QTNF5-related late-onset retinal degeneration. Retina. 2012;32(8):1643–51.
74. Jiao X, et al. Identification and population history of CYP4V2 mutations in patients with Bietti crystalline corneoretinal dystrophy. Eur J Hum Genet. 2017;25(4):461–71.
75. Li A, et al. Bietti crystalline corneoretinal dystrophy is caused by mutations in the novel gene CYP4V2. Am J Hum Genet. 2004;74(5):817–26.
76. Ng DS, et al. Genetics of Bietti crystalline dystrophy. Asia Pac J Ophthalmol (Phila). 2016;5(4):245–52.
77. Lee J, et al. The metabolism of fatty acids in human Bietti crystalline dystrophy. Invest Ophthalmol Vis Sci. 2001;42(8):1707–14.
78. Lee J, et al. Identification, isolation, and characterization of a 32-kDa fatty acid-binding protein missing from lymphocytes in humans with Bietti crystalline dystrophy (BCD). Mol Genet Metab. 1998;65(2):143–54.
79. Lai TY, et al. Genotype phenotype analysis of Bietti's crystalline dystrophy in patients with CYP4V2 mutations. Invest Ophthalmol Vis Sci. 2007;48(11):5212–20.
80. Bernauer W, Daicker B. Bietti's corneal-retinal dystrophy. A 16-year progression. Retina. 1992;12(1):18–20.
81. Vargas M, et al. Bietti crystalline dystrophy. In: Adam MP, et al., editors. GeneReviews (R). University of Washington, Seattle; 1993.
82. Fuerst NM, et al. Detailed functional and structural phenotype of Bietti crystalline dystrophy associated with mutations in CYP4V2 complicated by choroidal neovascularization. Ophthalmic Genet. 2016;37(4):445–52.
83. Akincioglu D, et al. Objective determination of retinal function in Bietti crystalline retinopathy. Turk J Ophthalmol. 2016;46(3):144–7.
84. Raoof N, Vincent AL. Novel gene mutation in a patient with Bietti crystalline dystrophy without corneal deposits. Clin Exp Ophthalmol. 2017;45(4):421–4.
85. Halford S, et al. Detailed phenotypic and genotypic characterization of bietti crystalline dystrophy. Ophthalmology. 2014;121(6):1174–84.
86. Rossi S, et al. Clinical and genetic features in Italian Bietti crystalline dystrophy patients. Br J Ophthalmol. 2013;97(2):174–9.
87. Manzouri B, et al. Bietti crystalline retinopathy: report of retinal crystal deposition in male adolescent siblings. Arch Ophthalmol. 2012;130(11):1470–3.
88. Okialda KA, et al. Bietti crystalline dystrophy. In: Pagon RA, et al., editors. GeneReviews(R). University of Washington, Seattle; 1993.
89. Parravano M, et al. Bietti crystalline dystrophy: a morpho-functional evaluation. Doc Ophthalmol. 2012;124(1):73–7.
90. Liu DN, et al. The characterization of functional disturbances in Chinese patients with Bietti's crystalline dystrophy at different fundus stages. Graefes Arch Clin Exp Ophthalmol. 2012;250(2):191–200.
91. Padhi TR, Kesarwani S, Jalali S. Bietti crystalline retinal dystrophy with subfoveal neurosensory detachment and congenital tortuosity of retinal vessels: case report. Doc Ophthalmol. 2011;122(3):199–206.
92. Sen P, Ray R, Ravi P. Electrophysiological findings in Bietti's crystalline dystrophy. Clin Exp Optom. 2011;94(3):302–8.
93. Rossi S, et al. An atypical form of Bietti crystalline dystrophy. Ophthalmic Genet. 2011;32(2):118–21.
94. Mansour AM, Uwaydat SH, Chan CC. Long-term follow-up in Bietti crystalline dystrophy. Eur J Ophthalmol. 2007;17(4):680–2.
95. Gekka T, et al. CYP4V2 mutations in two Japanese patients with Bietti's crystalline dystrophy. Ophthalmic Res. 2005;37(5):262–9.
96. Lockhart CM, et al. Longitudinal characterisation of function and structure of Bietti crystalline dystrophy: report on a novel homozygous mutation in CYP4V2. Br J Ophthalmol. 2018;102:187.

97. Tabatabaei A, et al. A case of Bietti crystalline dystrophy with preserved visual acuity and extinguished electroretinogram: a case report. Cases J. 2009;2:7100.
98. Simunovic MP, Moore AT. The cone dystrophies. Eye (Lond). 1998;12 (Pt 3b):553–65.
99. Michaelides M, Hunt DM, Moore AT. The cone dysfunction syndromes. Br J Ophthalmol. 2004;88(2):291–7.
100. Michaelides M, et al. Progressive cone and cone-rod dystrophies: phenotypes and underlying molecular genetic basis. Surv Ophthalmol. 2006;51(3):232–58.
101. Aleman TS, et al. Spinocerebellar ataxia type 7 (SCA7) shows a cone-rod dystrophy phenotype. Exp Eye Res. 2002;74(6):737–45.
102. Birtel J, et al. Clinical and genetic characteristics of 251 consecutive patients with macular and cone/cone-rod dystrophy. Sci Rep. 2018;8(1):4824.
103. Fishman GA, et al. ABCA4 gene sequence variations in patients with autosomal recessive cone-rod dystrophy. Arch Ophthalmol. 2003;121(6):851–5.
104. Birch DG, et al. Visual function in patients with cone-rod dystrophy (CRD) associated with mutations in the ABCA4(ABCR) gene. Exp Eye Res. 2001;73(6):877–86.
105. Demirci FY, et al. X-linked cone-rod dystrophy (locus COD1): identification of mutations in RPGR exon ORF15. Am J Hum Genet. 2002;70:1049–53.
106. Robson AG, et al. Functional correlates of fundus autofluorescence abnormalities in patients with RPGR or RIMS1 mutations causing cone or cone rod dystrophy. Br J Ophthalmol. 2008;92(1):95–102.
107. Branham K, et al. Mutations in RPGR and RP2 account for 15% of males with simplex retinal degenerative disease. Invest Ophthalmol Vis Sci. 2012;53(13):8232–7.
108. Sakuramoto H, et al. Two siblings with late-onset cone-rod dystrophy and no visible macular degeneration. Clin Ophthalmol. 2013;7:1703–11.
109. Malm E, et al. Full-field electroretinography and marked variability in clinical phenotype of Alstrom syndrome. Arch Ophthalmol. 2008;126(1):51–7.
110. Marshall JD, et al. Alstrom syndrome. Eur J Hum Genet. 2007;15(12):1193–202.

Chapter 7
Syndromic Disorders

Alfonso Senatore, Wajiha Jurdi Kheir, Minzhong Yu, Alessandro Racioppi, Roberto Gattegna, Donnell Creel, and Alessandro Iannaccone

Hearing Loss Syndromes

Usher Syndrome

Usher syndrome is an autosomal recessive disorder caused by a mutation in one of at least 11 genes that has both hearing loss and progressive retinal degeneration of the retinitis pigmentosa (RP) type. Usher syndrome accounts for about 50% of all deaf–blinding disorders. The exact prevalence of Usher syndrome in the population is not known, but it has been estimated to affect 1:10,000 to 1:20,000 people [1].

Usher syndrome is classified as type I, type II, and type III, mainly according to the age of onset and the severity of sensorineural hearing loss [2] (Table 7.1). Usher syndrome type I is typically characterized by profound congenital nonprogressive

A. Senatore · W. J. Kheir · A. Iannaccone
Center for Retinal Degenerations and Ophthalmic Genetic Diseases, Duke University School of Medicine, Duke Eye Center, Department of Ophthalmology, Durham, NC, USA

M. Yu (✉)
Department of Ophthalmology, University Hospitals Eye Institute, Cleveland, OH, USA
e-mail: minzhong.yu@uhhospitals.org

A. Racioppi
Center for Retinal Degenerations and Ophthalmic Genetic Diseases, Duke University School of Medicine, Duke Eye Center, Department of Ophthalmology, Durham, NC, USA

University of North Carolina, Chapel Hill, NC, USA

R. Gattegna
Center for Retinal Degenerations and Ophthalmic Genetic Diseases, Duke University School of Medicine, Duke Eye Center, Department of Ophthalmology, Durham, NC, USA

Retina Service, Israelitic Hospital, Rome, Italy

D. Creel
Moran Eye Center, University of Utah School of Medicine, Salt Lake City, UT, USA

M. Yu et al. (eds.), *Handbook of Clinical Electrophysiology of Vision*,
https://doi.org/10.1007/978-3-030-30417-1_7

Table 7.1 Clinical classification of the Usher syndromes

Clinical variables	Usher syndrome Type I	Usher syndrome Type II	Usher syndrome Type III	Pseudo-Usher syndrome
Inheritance	Autosomal recessive	Autosomal recessive	Autosomal recessive	X-linked recessive (*RPGR* gene)
Retinal disease	RP	RP	RP	RP
Onset	*Within second decade*	*Variable (as late as fourth or fifth decade)*	*Within second decade*	*M: w/in second decade* *F: Very variable*
Hearing loss	Sensorineural	Sensorineural	Sensorineural	Variable
Onset	*From birth*	*From birth*	*Late-onset*	*Variable*
Severity	*Moderate to profound*	*Mild to moderate*	*Mild to moderate*	*Mild to moderate*
Progressive	*No*	*No*	*Yes*	*Yes?*
Speech abnormality	Moderate to severe	Mild to moderate	Absent	Absent to moderate
Vestibular dysfunction	Present	Absent or minimal	Variable	Unknown (not studied yet)
Recurrent respiratory tract infections	No	No	No	Yes (variable)

hearing loss requiring cochlear implantation to avoid severe if not complete impediment of speech development and is caused by mutations in at least six genes. The most commonly involved gene is the *MYO7A* gene, followed by *CDH23* gene. Hearing loss in Usher syndrome type I is usually associated with balance problems due to vestibular dysfunction. Usher syndrome type II is characterized by mild to moderate congenital "sloping" hearing loss at the high frequencies and is caused by mutations in at least three genes, the most common being the *USH2A* gene. Hearing loss in Usher syndrome type II is usually nonprogressive, but deterioration over time can occur. Patients with Usher syndrome type II also show variable degrees of vestibular dysfunction that is not always present. Unlike Usher syndrome type I and II, Usher syndrome type III is characterized by mild to moderate postverbal progressive hearing loss and is usually caused by the mutations in the *CLRN1* gene. Because of the postverbal onset of the hearing loss, speech impediment is usually not seen in Usher syndrome type III. The extent of vestibular dysfunction is also variable. Types I and II are by far the most common ones, accounting for over 90% of the cases of Usher syndrome.

A type IV Usher syndrome with X-linked recessive inheritance was suspected to exist in the past, but this syndrome has now been shown to be a pseudo-Usher syndrome linked to mutations affecting the RRC1-like domain of the *RPGR* gene, a common cause of X-linked RP also expressed in the epithelial lining of the respiratory tract and in the inner ear. This variant is also associated with recurrent upper respiratory tract infections (especially otitis and sinusitis) and immotile cilia-like symptoms [3, 4].

The overall characteristics of the RP seen in Usher syndrome are illustrated in the earlier non-syndromic RP section of this book, and, to this date, they are not known to differ significantly from it (Fig. 7.1). In general, patients with Usher syndrome type I tend to exhibit visual symptoms within the first two decades of life and tend to be affected with an overall more severe form of RP [5, 6]. The onset of RP in Usher syndrome type II is much more variable; many patients are symptomatic from the first decade of life to patients who do not experience symptoms and signs of RP well into their fifth decade of life [7, 8]. Many patients with Usher syndrome type II may not exhibit the typical clinical fundus changes of RP for decades, even when symptoms may have already begun, whereby the diagnosis of Usher syndrome type II can be further delayed by this misleading paucity of clinical findings. The RP of Usher syndrome III is more similar in characteristics and age of onset to Usher syndrome type II, with initial symptoms usually within the first 2–3 decades of life. The progression of RP in Usher syndrome III has also been estimated to be more rapid than in Usher syndrome II [9].

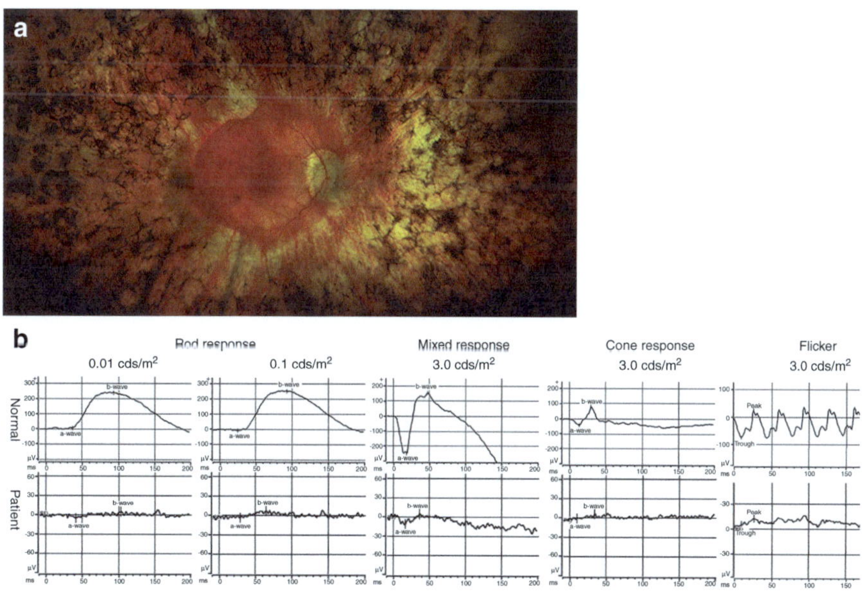

Fig. 7.1 Findings in Usher syndrome. (**a**) Color fundus photograph of a 76-year-old Caucasian patient with Usher syndrome type I due to homozygosity for a complex deletion disrupting the splicing acceptor site of intron 46 of the *CDH23* gene. Dense bone spicule-like pigment deposits affecting every retinal quadrant, attenuated vessels, and waxy disc pallor with peripapillary atrophy are visible. At this stage of the disease, the macula of this eye is still fairly healthy, with 20/40 acuity. (**b**) Full-field flash ERGs show non-recordable rod-driven responses (but detectable at slightly brighter flash intensities), markedly reduced but measurable mixed responses, and residual cone-driven responses. The ensemble of these findings is compatible with an advanced rod–cone pattern of retinal disease of the retinitis pigmentosa (RP) family

Refsum Disease

Refsum disease is also known as *heredopathia atactica polyneuritiformis* [10] or phytanic acid storage disease exhibiting some of the key clinical and metabolic features of this syndromic form of RP. Refsum disease is an autosomal recessive metabolic disorder due to peroxisomal dysfunction leading invariably to marked accumulations of circulating phytanic acid, a branched very long fatty acid. There are two main types of Refsum disease, the adult-onset form and the infantile-onset form (Table 7.2). Cardinal features of Refsum disease include pigmentary retinal degeneration (rod–cone, RP type), progressive sensorineural hearing loss, and a peripheral polyneuropathy. Furthermore, other specific differential features distinguishing the adult-onset from the infantile-onset form are also summarized in Table 7.2.

From a genetic point of view, there are at least two genes (*PHYH* and *PEX7*) causing the adult-onset form and two others (*PEX1* and *PEX2*) causing the infantile-onset form of Refsum disease. All these genes are implicated in peroxisomal function, the dysfunction that is the underlying common etiology of all forms of Refsum disease [11].

The clinical and functional signs of the retinal degeneration in Refsum disease are a rod–cone, RP-like retinopathy, and the visual electrophysiological findings are comparable to those of non-syndromic RP illustrated earlier in the book. The main differential diagnosis of Refsum disease is with the Usher and pseudo-Usher syndromes reviewed and mitochondrial conditions in which hearing loss is also

Table 7.2 Spectrum of clinical manifestations in Refsum disease

Main invariant features of Refsum disease	
Pigmentary retinal degeneration (rod–cone, RP type)	
Progressive sensorineural hearing loss[a, b]	
Peripheral polyneuropathy	
Elevated serum level of phytanic acid	
Adult-onset form	*Infantile-onset form*
[a]Postverbal onset, no speech impediment	[b]Early-onset, variable speech impediment
Main features	Mental retardation
Ataxia	Developmental delay
Cardiopathy	Hypotonia
Ichthyosis	Failure to thrive
Skeletal abnormalities:	Facial dysmorphism
Multiple epiphyseal dysplasia	Hepatomegaly
Bilateral fourth metatarsal shortening	Bile metabolism compromise with steatorrhea
	Osteoporosis
Accessory features	Serology:
Anosmia	Hypocholesterolemia + elevated levels of:
Ptosis	Very long chain fatty acids
Miosis	Di- and tri-hydroxycholestanoic acid
	Pipecolic acid

manifest, mainly the neuropathy–ataxia–retinitis pigmentosa (NARP) syndrome and the Kearns–Sayre syndrome (see below). The conclusive diagnosis of this particular condition does not rest of the results on the eye examination or the full-field electroretinography (ffERG) results as much as it does instead on serological testing for phytanic acid levels.

Phytanic acid accumulates not only in the blood stream but also in the target tissues affected by the disease, whereby early recognition of Refsum disease is essential to mitigate its severity, as a dietary regimen low in butter, animal fat, and other foods rich in phytanic acid and plasmapheresis, performed once or twice a month, has been reported to stabilize disease progression, at least systemically [12].

Wolfram Syndrome (DIDMOAD)

Wolfram syndrome is a rare autosomal recessive disease, also known with the acronym DIDMOAD, that encapsulates the typical features of the full-fledged phenotypic expression of the disorder, namely, diabetes insipidus (DI), insulin-dependent diabetes mellitus (DM), progressive optic atrophy (OA), and deafness (D, in the form of progressive sensorineural hearing loss). Optic atrophy and diabetes mellitus are typically the earliest and, often, the only signs of the disease typically observed from the first or second decade of life.

Two causal genes for Wolfram syndrome have been cloned to date: *WFS1* [13, 14] encoding for wolframin, a transmembrane glycoprotein of the endoplasmic reticulum, and *WFS2* or *CISD2* [15, 16] encoding an endoplasmic reticulum intermembrane small protein (ERIS) localized to the endoplasmic reticulum as well. The expression patterns of wolframin have been extensively characterized [17, 18].

Patients with this rare disorder can be also affected with a pigmentary retinal dystrophy and appear to be exquisitely prone to developing also neuropsychiatric disorders, as are the heterozygous carriers of the trait. In addition to diabetic complications, affected patients are at risk for premature death from progressive brainstem atrophy and complications of urinary tract atony. Vitamin B1 (thiamine) supplementation may reduce the insulin needs of these patients, but it is unknown if this may be of benefit also for the other features of the disease [19].

Typically, from a visual electrophysiological standpoint, patients with Wolfram syndrome will exhibit various degrees of visual evoked potential (VEP) amplitude reduction, usually proportional to the severity of the optic atrophy. The ffERG will show abnormalities, usually of the rod–cone, RP-like type, if a pigmentary retinal dystrophy is also observed.

Mitochondrial Diseases

Kearns–Sayre Syndrome

Kearns–Sayre syndrome (KSS) is a complex mitochondrial condition caused by spontaneous large deletions of mitochondrial DNA (mtDNA), whereby the traditional maternal lineage inheritance pattern of mitochondrial diseases is typically absent. KSS overlaps significantly with both Usher syndrome and Refsum disease and, thus, must be distinguished from both conditions. Differential diagnostic features are not seen in either one of these conditions but instead found only in KSS including (Fig. 7.2) retinal degeneration of the cone–rod type (CORD) with atrophy of the RPE at the posterior pole and around the disc, whereby KSS patients usually complain of light aversion more so than night blindness, and their central vision is usually affected more than their peripheral vision, bilateral chronic progressive external ophthalmoplegia (CPEO) with ptosis, and heart conduction problems, making these patients susceptible to sudden heart block. The symptoms include ragged red fibers (RRF) of the striate muscles at the time of muscle biopsy and lactic

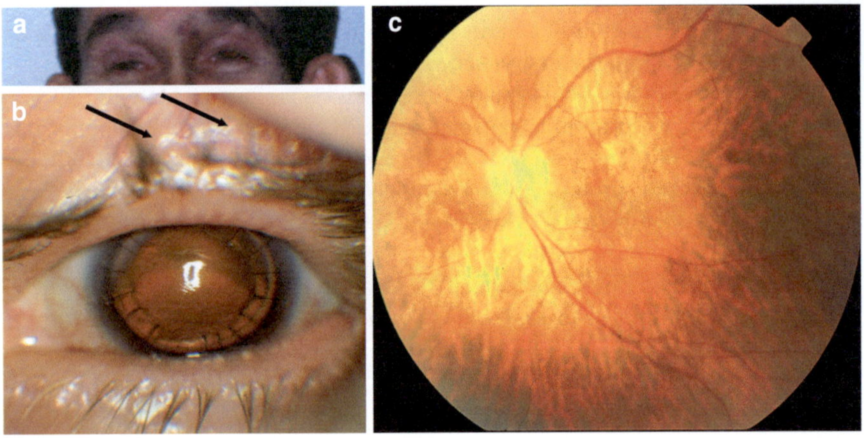

Fig. 7.2 Findings in Kearns–Sayre Syndrome (KSS). The images are from a 50-year-old Caucasian man whose review of systems was positive for severe progressive sensorineural hearing loss with speech impediment since age 7, infantile seizures, "irregular heartbeat," migraines, short stature, and a recent history of diabetes mellitus with progressive weight loss. He reported mildly progressive deterioration of vision since age 7 and progressive ptosis (**a**) since his mid-20s, requiring surgical correction (see arrows in **b**) and ophthalmoplegia. Marked worsening of visual acuity over the previous 3 years occurred due to corneal clouding requiring a penetrating keratoplasty that, however, did not lead to significant visual improvement. (**c**) Fundus photography reveals a large area of partial RPE atrophy involving the posterior pole and the peripapillary region and disc pallor. The ffERG of this patient (not shown) exhibited a cone–rod pattern of retinal dysfunction, with moderate rod-driven and mixed response reduction, and more severe reduction of the cone-driven electroretinography (ERG) responses, associated with large central scotomas to visual field testing. Serum lactate and pyruvate were elevated, and a muscle biopsy confirmed a large mtDNA deletion, establishing the diagnosis of KSS. Two years after the diagnosis, this patient developed sudden heart block and he received a pacemaker

acidosis, recognized by the identification of elevated serum lactate and pyruvate levels. Prophylactic pacemaker implantation is recommended and can save the life of affected patients.

Other manifestations characterizing KSS include deafness, diabetes mellitus, muscle weakness in the limbs, kidney problems, developmental delay, short stature in conjunction with growth hormone deficiency, ataxia, seizure disorders, hypothyroidism, and delayed sexual maturation [20, 21]. More rarely, corneal opacities can develop due to corneal endothelial dysfunction [22], requiring in severe cases a corneal transplant (Fig. 7.2). However, in recent years, it has been shown coenzyme Q-10 supplementation can ameliorate corneal involvement [23, 24]. A similar beneficial impact has been observed also on CPEO and other manifestations of the disease. The genetic diagnosis of KSS can usually be verified only by mtDNA analysis on muscle biopsy tissue specimens.

The visual electrophysiological characteristics of KSS include the typical of CORD, with cone > rod response amplitudes that vary in severity [21]. A classification in various disease stages based on the severity of the retinal disease has been also proposed [25].

Neuropathy–Ataxia–Retinitis Pigmentosa Syndrome

The neuropathy–ataxia–retinitis pigmentosa (NARP) syndrome is a rare disorder with mitochondrial inheritance caused by the T8993G point mtDNA mutation in the MTATP6 gene, encoding the subunit 6 of the mitochondrial H(+)-ATPase [26], usually best detected via mtDNA testing performed on muscle biopsy tissue. Unlike KSS, NARP is usually maternally transmitted, but mothers carrying the T8993G mutation are usually asymptomatic.

The onset of the NARP syndrome is usually in childhood or early adulthood, including night blindness, photophobia, and peripheral visual field loss. These symptoms herald the development of a fairly severe RP-like retinopathy, despite the common lack of diffuse bone spicule deposits and presence instead of a salt and pepper retinopathy with coarse RPE mottling. This retinopathy is associated with severe ffERG response depression that usually exhibits a rod–cone pattern of degeneration, followed by non-recordable responses usually by the third decade of life. However, a cone–rod pattern has also been reported in some NARP patients [27]. The picture is often complicated further by optic atrophy and ophthalmoplegia.

In addition to these ophthalmic manifestations, NARP is associated with a number of systemic findings, such as peripheral neuropathy, ataxia in conjunction with corticospinal tract atrophy, hearing loss, short stature, proximal muscle weakness, learning difficulties, seizures, sleep apnea, and/or cardiac arrhythmias [28]. Recently, progressive kidney function compromise leading up to failure and requiring dialysis has also been reported [29]. These systemic findings can vary significantly in severity and age of onset, making the diagnosis of NARP (vs. non-syndromic RP) potentially challenging, and the condition can be confused with both Usher syndrome and Refsum disease.

Maternally Inherited Diabetes and Deafness Syndrome

The maternally inherited diabetes and deafness (MIDD) syndrome is a condition characterized by the constellation of insulin-dependent diabetes mellitus and hearing loss with onset in the mid adulthood. MIDD is caused by mitochondrial DNA mutations that in 85% of the cases are the A3243G point mutation in the *MTTL1* gene [30]. Its symptoms typically include bilateral macular pattern dystrophy (MPD) existing in as many as 85% of the cases and more rarely external ophthalmoplegia, ptosis, cardiomyopathy, myopathy, renal problems, and neuropsychiatric symptoms [30–32].

The type, extent, and severity of macular involvement vary between patients and within families [33–35]. The progression of the MIDD–MPD syndrome is relatively slow [36, 37], and, in its earlier stages, it can remain asymptomatic for years until functional loss around the fovea develop, causing reading difficulties even when acuity may still be 20/20. In more advanced stages, MIDD patients can develop frank macular atrophic areas resembling geographic atrophy and may progress to involve the fovea, leading to severe decrease in central vision (Fig. 7.3). Importantly, despite

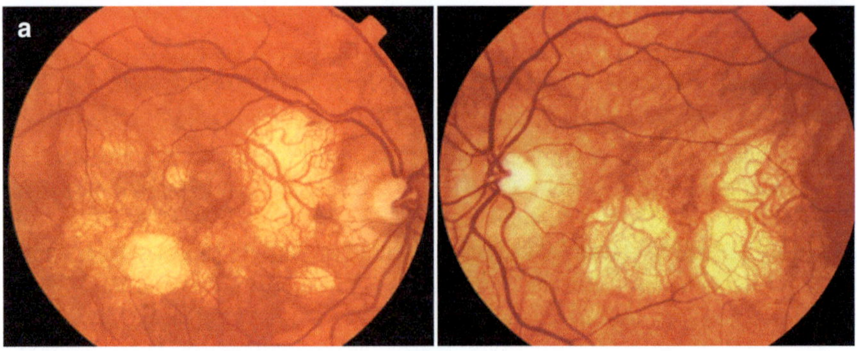

Fig. 7.3 Fundus features of the maternally inherited diabetes and deafness with macular pattern dystrophy (MIDD–MPD) syndrome. (**a**) Patient 1: Color fundus photographs from a 49-year-old Caucasian woman with a 6-year history of light aversion and a 3-year complaint of blurred near vision and pericentral scotomas despite 20/20 acuity in both eyes. This patient also has sloping sensorineural hearing loss, a 5-year history of insulin-dependent diabetes mellitus, a 3-year history of nonalcoholic liver cirrhosis, and a 5-year history of bilateral acral peripheral neuropathy affecting both feet. Maternal history of hearing loss, headaches, seizures, and developmental delay was documented. Note the bilateral multifocal atrophic perifoveal lesions at the posterior pole. (**b**) Late phase fluorescein angiograms show, in addition to the large window defects due to the atrophic lesions, the diagnostic finding supporting the diagnosis of macular pattern dystrophy (MPD) of linear, punctate, hypofluorescent lesions disposed in a spoke-wheel pattern around the macula and between and outside of the patches of atrophy. (**c**) Patient 2: Color fundus photographs from a Caucasian 44-year-old man with light-to-dark adaptation problems, trouble reading at near, and constant image of "snowy TV screen" around fixation since his late 30s, when he was found to have macular atrophy. Insulin-dependent diabetes mellitus (type 1) without evidence of retinopathy and progressive sensorineural hearing loss had been diagnosed since age 32. The macular region shows multiple patches of RPE atrophy sparing the fovea in the right eye. In the left eye, the atrophy is more severe, and a yellowish central fibrotic lesion due to a neovascular complication occurred 1 year earlier can be seen. (**d**) Fundus autofluorescence images show a large central area of patchy hypo-autofluorescence due to RPE loss, surrounded by radially shaped hyper-autofluorescent patterns, diagnostic for MPD. Both patients bear the molecularly confirmed diagnosis of MIDD–MPD due to the common A3243G $tRNA^{Leu}$ mtDNA mutation, confirmed on muscle tissue after a muscle biopsy had been obtained, following evidence for elevated serum lactate and pyruvate

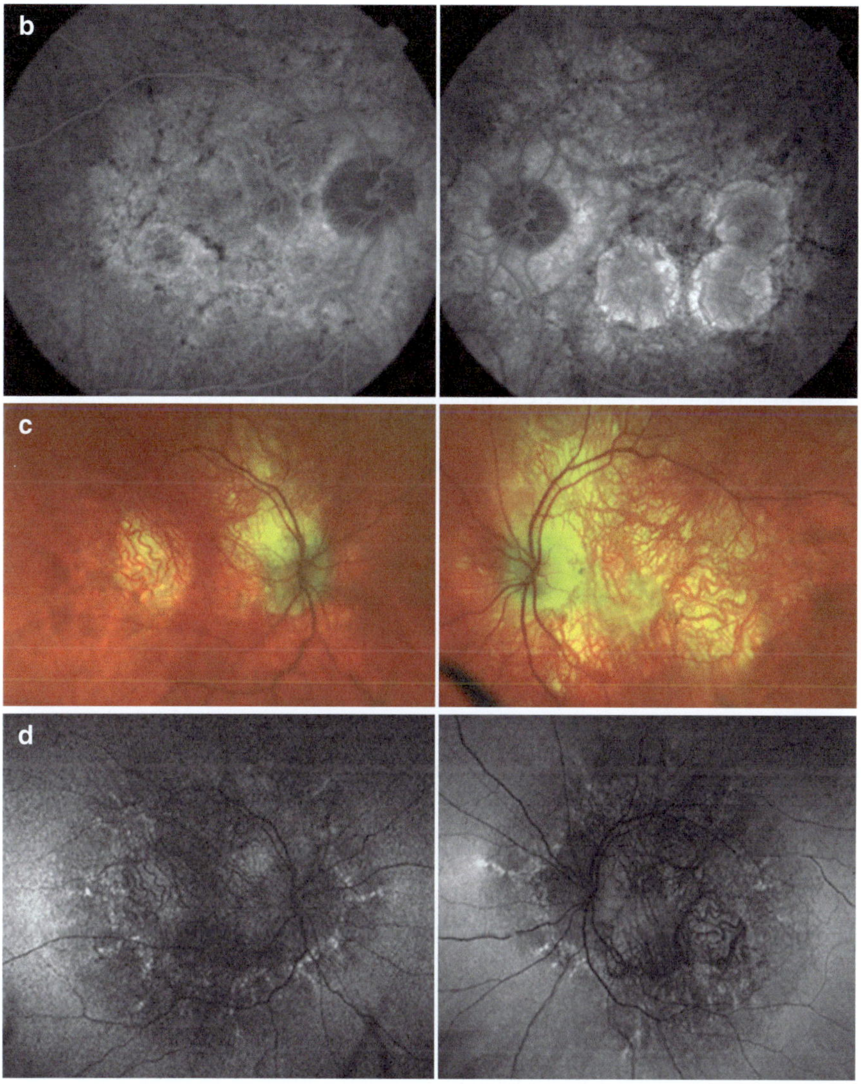

Fig. 7.3 (continued)

the macular changes, these diabetic patients hardly ever develop diabetic retinopathy changes [33]. In some cases, neovascular complications, mimicking in every way AMD, can also occur, as it was the case of the patient shown in Fig. 7.3c, d who had neovascular complications in the left eye, making the differential diagnosis with AMD even more challenging. The triad of insulin-dependent diabetes mellitus, hearing loss, and macular dystrophy, however, is usually seen by ages 40–50, an unusually early age of onset for AMD. Thus, whether associated with overt maternal history of diabetes or not, this triad should invariably prompt as a minimum workup for serum lactate and pyruvate levels usually elevated in MIDD–MPD syndrome patients, and then subsequent verification of the diagnosis via mitochondrial DNA diagnostic testing.

From the visual electrophysiological standpoint, the ffERG is usually not a sensitive diagnostic test in MIDD–MPD patients, although, in more advanced stages, patients can show abnormal rod and cone responses. The multifocal electroretinogram (mfERG) usually shows early with decreased response amplitudes, corresponding to the areas more affected by the retinal disease process (Fig. 7.4). In addition, many MIDD–MPD patients also exhibit reduced pattern electroretinogram (PERG) P50 amplitudes [35, 38]. The EOG can also be abnormal, as it was the case in the patient illustrated in Fig. 7.3a, b, reflecting a widespread RPE abnormality in these patients in keeping with the frequent RPE abnormalities that most mitochondrial retinopathies tend to exhibit. It is important to clarify neither one of these visual electrophysiological tests is diagnostic for the MIDD–MPD syndrome, and in addition to the noted pathognomonic clinical triad, from the ophthalmological standpoint, fluorescein angiography, and fundus autofluorescence are typically diagnostic for the linear spoke–wheel RPE macular changes characterizing the MPD of the MIDD–MPD syndromic association (Fig. 7.3) and are essential in confirming the suspicion of this condition.

Recognition of this entity has important treatment implications because it has been reported the insulin need of these diabetic patients are significantly reduced while on pro-mitochondrial dietary regimens such as coenzyme Q-10 [33] and, in our experience, can extend also to the macular phenotype.

Obesity Syndromes

Bardet–Biedl Syndrome

Bardet–Biedl syndrome (BBS), first described by Bardet in 1920 and Biedl in 1922, is an autosomal recessive disease with a wide spectrum of clinical systemic features, including congenital postaxial abnormalities of the extremities (polydactyly, syndactyly, brachydactyly, and/or clinodactyly of the fifth finger), early onset obesity, a wide range of renal abnormalities and dysfunction leading up, in more severe cases, to kidney failure, abnormalities of the secondary sexual development, dental anomalies, and a wide range of smell sensory loss [39, 40]. Once believed to be a cardinal feature of BBS, developmental delay can often be absent, although behavioral abnormalities up to frank autistic features are usually manifest [41]. More rarely, BBS is associated also with diabetes mellitus, other endocrinological disturbances, and hepatic fibrosis [39, 42]. Over 20 genes responsible for BBS have been identified. These genes encode proteins involved in the development and the maintenance of the primary cilium, the so-called BBSome [43].

The ophthalmic manifestations of BBS are typically characterized by a rod–cone dystrophy pattern of the RP type, although commonly associated with delayed clinically visible features of the disease that can significantly delay the diagnosis [44–46]. Under typical circumstances, severe rod–cone patterns of ffERG dysfunction

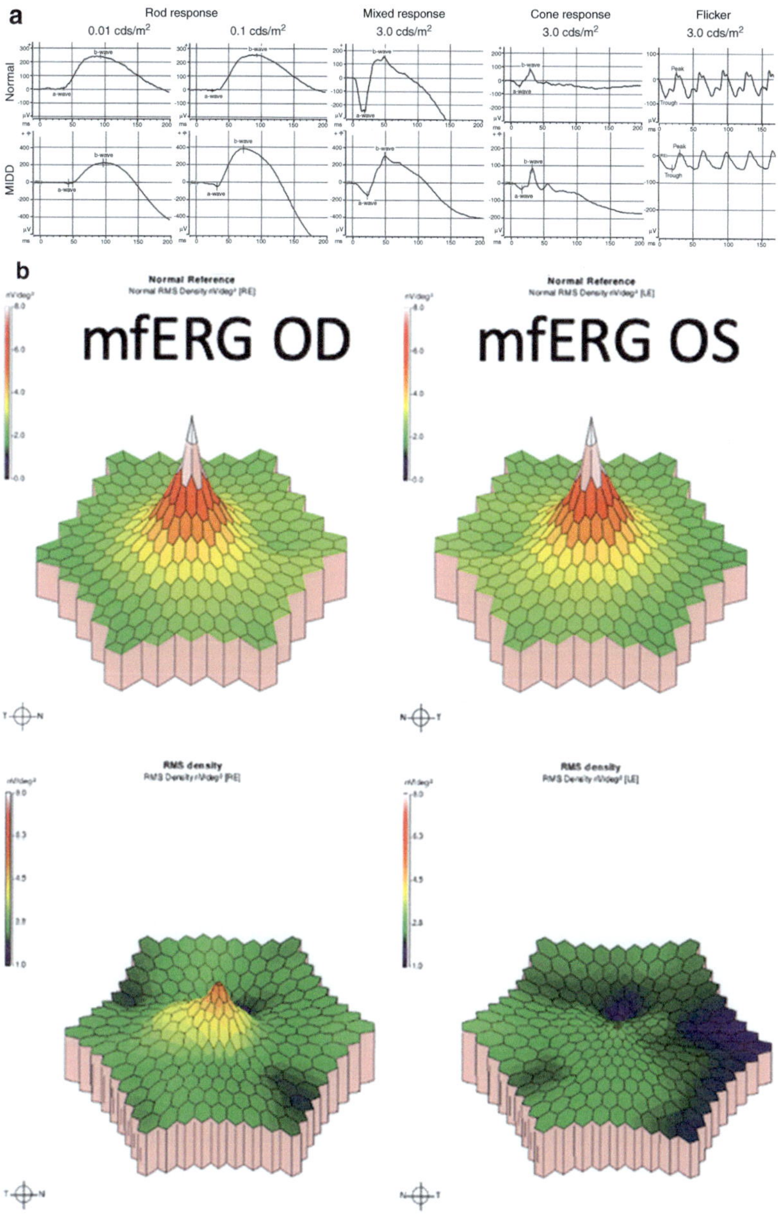

Fig. 7.4 Electrophysiologic findings in the MIDD–MPD syndrome. (**a**) The ffERG of Patient 2 (from Fig. 7.3) shows normal rod and transient cone responses: The mixed response a-wave is reduced, and both the mixed b-wave and the flicker response are borderline low in amplitude. (**b**) The mfERG shows a patchy asymmetric response depression and significant reduction of the central peak in OS but not in OD, consistent with the pattern of fundus abnormalities shown in Fig. 7.3. The ffERG of Patient 1 (not shown) was normal in amplitude, but the cone ERGs were delayed. In this patient, an EOG was also measured (not shown) that was frankly pathological bilaterally, with an Arden ratio of 1.35, diagnostic for widespread RPE dysfunction

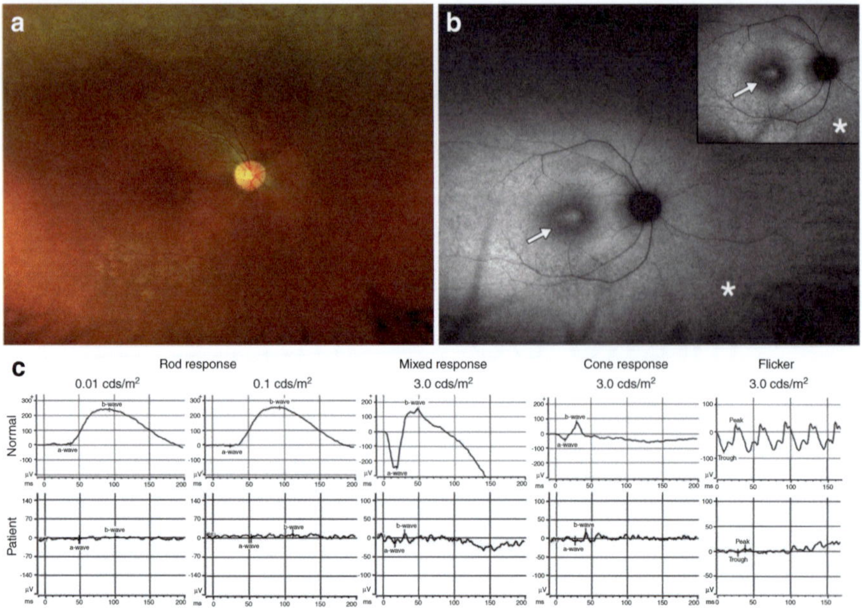

Fig. 7.5 Findings in Bardet–Biedl syndrome (BBS). (**a**) Color fundus photograph of a 16-year-old Middle-Eastern female BBS patient homozygous for a splice site *BBS2* mutation, with vision loss, diet-responsive obesity (BMI > 29), polydactyly and brachydactyly of both hands and feet, and developmental delay. There is mild waxy disc pallor of the nerve, a beaten bronze appearance at the macula, vascular attenuation, and salt and pepper peripheral appearance without visible bone spicule-like deposits. (**b**) Fundus autofluorescence from the same patient reveals much more overtly the extent of the macular involvement, with a bull's eye-shaped perifoveal hypo-autofluorescent ring attributable to RPE loss (arrows), a typical finding for BBS patients highlighted further in the inset, and a speckled hypo-autofluorescent mid-peripheral pattern corresponding to the noted salt and pepper changes (asterisks). (**c**) The ffERG of the same patient shows dark-adapted rod-driven and mixed recordings that are non-recordable to dark-adapted rod-driven stimuli and markedly diminished residual responses to all other stimuli

will be observed even when the typical findings of RP in fundus appearance is still missing (Fig. 7.5). This is in line with the underlying ciliopathy etiology of BBS, whereby the clinical appearance can be deceptively normal or near normal, while outer segment dysfunction and loss may already be underway, consistent with the observation that the ffERG in BBS tends to be far more frequently non-recordable than non-syndromic RP [44]. However, cases of initially normal or near-normal ffERG have been documented in molecularly confirmed cases of BBS [44, 46]. Furthermore, cases with a cone–rod phenotype matching best the Lawrence–Moon spectrum of the disease have also been reported [44].

Since ciliopathies affect typically both rods and cones, macular damage and dysfunction is extremely common in BBS, whereby the mfERG response tends to be consistently abnormal, with or without foveal peak sparing, depending on the severity of the foveal involvement [47].

VEP testing can be informative about the health of the optic nerve in BBS patients [48], that is more frequently abnormal than in non-syndromic RP [45], although the interpretation of VEP abnormalities in BBS can be challenging. Delayed and reduced flash and pattern VEP responses are very common in BBS, in part contributed by the intrinsically defective macular photoreceptor function and in part by a variety of optic abnormalities ranging from marked pallor/atrophy to swelling [45, 46, 48].

Alström Syndrome

Alström syndrome is a rare autosomal recessive disease, first described in 1949, with a broad spectrum of clinical symptoms, largely overlapping with BBS. To date only one causal gene has been mapped and cloned, the *ALMS1* gene [49–51], with approximately 950 reported cases in literature [52]. The ALMS protein is expressed at the level of the centrosomes and the basal bodies of ciliated cells, suggesting the role in microtubule organizing to assemble the mitotic spindle and forming the microtubule scaffold guiding intracellular organelle and vesicle trafficking and in anchoring the microtubular axoneme of cilia and flagella to help in the organization of intraflagellar transport. This shared ciliopathy mechanism explains well the similarity between Alström syndrome and BBS. A mouse model recapitulating fairly well the human phenotype has also been created [53].

From a systemic standpoint, this obesity syndrome is characterized by infantile-onset dilated cardiomyopathy, postverbal onset progressive sensorineural hearing loss, and delayed-onset diabetes mellitus (usually preceded by hyperinsulinemia), in the absence of polydactyly or other dystrophic aspects of the extremities (typically seen in BBS instead). Renal, hepatic, and pulmonary failure due to fibrosis can also occur and be the causes for a poor prognosis in Alström syndrome patients. A French-Acadian variant of Alström syndrome, found in Louisiana, has also been associated with developmental delay otherwise typically absent (Fig. 7.6).

For all patients, an early-onset retinopathy with markedly reduced acuity, photophobia, nystagmus, and a bull's eye maculopathy appearance, is usually the main ophthalmic finding of this rare syndrome. The ffERG shows both rod and cone responses decrease considerably in Alström syndrome in a severe, progressive cone–rod dystrophy pattern. In longitudinal studies, it has been also demonstrated the decrease of cone-driven responses usually precedes the decrease of rod-driven responses, indicating this condition follows a cone–rod pattern of retinal degeneration longitudinally [54–62].

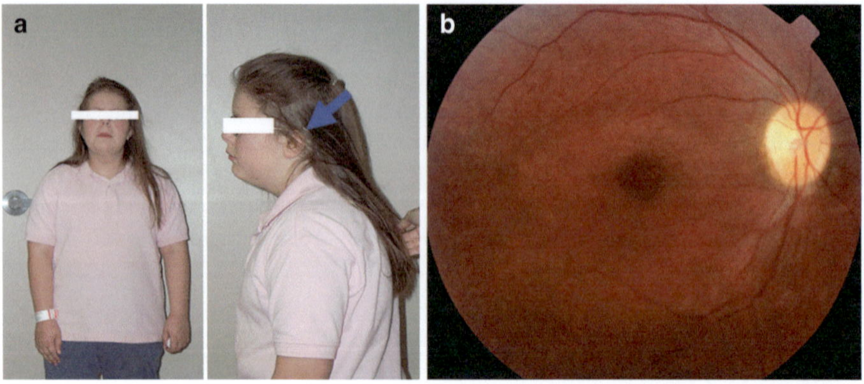

Fig. 7.6 Systemic and fundus findings in Alström syndrome. (**a**) Images of a 10-year-old overweight French-Acadian female patient with nystagmus, early-onset poor visual acuity, and light aversion. Systemically she displayed infantile-onset dilated cardiomyopathy, mild obesity, developmental delay, progressive postverbal sensorineural hear loss requiring the use of a hearing aid (white arrow), and recent onset insulin-dependent diabetes. (**b**) Color fundus photography shows a perifoveal bull's eye macular appearance, enhanced vitreoretinal interface reflexes, mild vascular attenuation, and temporal disc pallor. No pigmentary deposits were observed. The ffERG of this patient (not shown) exhibited a marked decrease in rod (5–10% of normal values) and mixed responses (15% of normal values), with a barely recordable cone-driven ERG, consistent with a severe cone–rod dystrophy pattern. The ensemble of these findings is diagnostic for Alström syndrome

Cohen Syndrome

Cohen syndrome is an autosomal recessive disorder with multiple congenital anomalies and mental retardation caused by the mutation of *COH1* (human homologue of Saccharomyces cerevisiae *VPS13B*) gene [63, 64]. This condition was first reported by Cohen in 1973 and is more common in Finland and among the Ashkenazi Jews and the Amish, with over 100 cases reported to date in literature [63–68].

COH1 is a very large gene and yields several different transcripts differentially expressed in various tissues and have complex domain structure. All protein isoforms are thought to function in vesicle-mediated transport and sorting of proteins within the cell and to play a role in the development and the function of the eye. Of note, the vast majority of *COH1* mutations found in Cohen syndrome patients are duplications [69], posing a significant diagnostic challenge if microarray-based approaches or effective in-silico copy number variant (CNV) assessment of sequencing data is not used [70].

The following are considered the minimal diagnostic criteria for Cohen syndrome (Table 7.3): (1) characteristic facial gestalt (short philtrum with grimacing

Table 7.3 Diagnostic criteria for Cohen syndrome

Systemic/laboratory
Facial *gestalt*[a] of:
Short philtrum with grimacing expression on smiling
Thick hair with low-set hairline, thick eyebrows
Prominent nose (often beak-shaped)
Microcephaly[b] (cases with macrocephaly have been reported)
Joint hyperextensibility[b]
Truncal childhood-onset obesity with slender extremities[b]
Nonprogressive developmental delay with cheerful disposition
Short stature (cases with tall stature have been reported)
Long hands with tapering fingers
Intermittent neutropenia[a]
Mild to moderate, not life-threatening
Ophthalmological
Down-slanting palpebral fissures with wave-shaped eyelids and thick eyelashes[a]
Chorioretinal dystrophy of the rod–cone type[a] (absent in the Ashkenazi Jews?)
Early-onset, progressive myopia[b]

[a]Minimal diagnostic criteria set forth by Chandler et al. 2003
[b]Most common ancillary diagnostic findings in molecularly confirmed cases of Cohen syndrome as per Chandler et al. 2003

expression on smiling and with thick hair with low-set hair line, thick eyebrows, and prominent nose); (2) intermittent neutropenia (usually ranging from mild to moderate and not life-threatening); (3) down-slanting palpebral fissures with wave-shaped eyelids and thick eyelashes; and (4) chorioretinal dystrophy of the rod–cone type (reportedly absent in patients of Ashkenazi-Jewish descent). Other ancillary ocular and systemic findings are summarized in Table 7.3. While microcephaly and short stature are typically observed, cases with macrocephaly and tall stature have also been reported.

From a retinal and visual electrophysiological standpoint, Cohen syndrome patients usually display an RP-like chorioretinal dystrophy of the rod–cone type (Fig. 7.7), *associated* with an early-onset, progressive myopia. Studies performed on younger patients demonstrate abnormal retinal findings, and ERG changes are early signs in the Cohen syndrome. Attenuation of all ffERG responses is progressive and can range from markedly attenuated to non-recordable [71]. As noted above, this RP-like retinopathy is usually absent in the Ashkenazi Jews [63, 64, 71, 72].

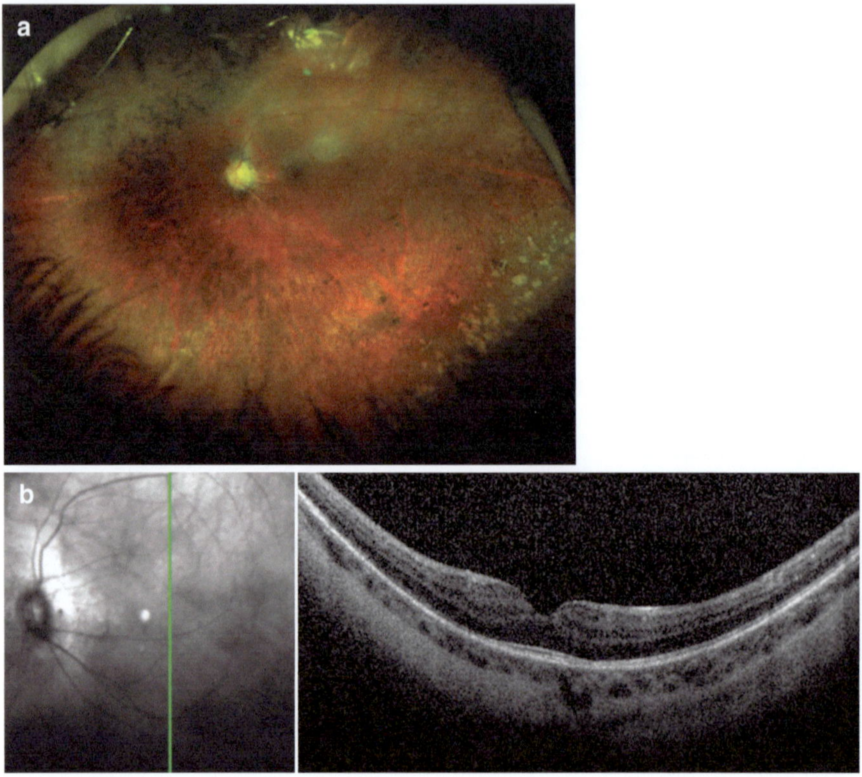

Fig. 7.7 Fundus and imaging findings in Cohen syndrome. (**a**) Color fundus photograph of a 23-year-old Middle-Eastern male patient homozygous for a large novel duplication of the *VPS13B (COH1)* gene encompassing exons 40–43 with night blindness as a chief visual complaint of 2–3 years, showing peripheral bone spicule-like pigmentary deposits, paving stonelike degeneration in the inferotemporal quadrant, waxy disc pallor, vascular attenuation, and a mild myopic macular staphyloma with an otherwise fairly healthy macula. (**b**) Vertical scan of the macular OCT showing excellent foveal EZ preservation, with mild surrounding retinal thinning and mild ONL attenuation. The systemic conditions precluded obtaining a ffERG, which would be predicted to exhibit a rod–cone pattern of retinal dysfunction of the RP type. His review of systems was remarkable for having syndromic facies characterized by micrognathia, microcephaly, low-set large ears, beak-shaped nose, thick philtrum, and abnormal teeth, developmental delay with very limited verbal skills, initially attributed to cerebral palsy, slender fingers (arachnodactyly-like), and a very wide stance with tarsus varus. His sister had a very similar facial gestalt and a more severe array of systemic symptoms

Other Syndromes

Cockayne Syndrome

Cockayne's syndrome is a rare autosomal recessive disorder, first reported by Cockayne in 1936, caused by mutation in several distinct genes appertaining at the excision-repair cross-complementing groups, *ERCC6* and *ERCC8* [73, 74]. This syndrome shows segmental premature aging due to a deficiency in transcription and DNA repair mechanisms and shares pathogenetic mechanisms with xeroderma pigmentosum. The main cellular defect is linked to the cellular impossibility in repairing oxidation-induced damage to DNA bases and the slow recovery of RNA and DNA synthesis after UV irradiation [75, 76]. The diagnosis is based on clinical assessment and lab DNA repair tests on fibroblast cell cultures from skin biopsies. Two main types were described: type I (classical type) with postnatal onset and type II (severe type) with prenatal onset, resulting in death by age of 6–7.

Cockayne's syndrome is characterized by deterioration in multisystem with cachectic dwarfism and growth deficiency, resulting in a precociously senile appearance. Sensorineural hearing loss, skin photosensitivity (as in xeroderma pigmentosum), mental retardation, marble epiphyses in some digits with disproportionately long limbs with large hands and feet, and flexion contractures of joints (knee contractures result in a "horse-riding" stance) are all clinical features of this severe syndrome. Hypertension and renal disease are frequent complications, leading both to an overall diminished life expectancy.

Ocular features of Cockayne's syndrome include an RP-like (rod–cone dystrophy type) pigmentary retina degeneration, with a salt and pepper appearance initially and then progressing to severe degeneration thereafter, optic atrophy (possibly secondary in nature), nystagmus, and cataracts (with difficult surgery due to miotic pupils and high hyperopic, *quasi*-microphthalmic refractive defect). Furthermore, patients can have corneal and eyelid features of xeroderma pigmentosum with decreased lacrimation and dry eye syndrome, conjunctival hyperemia, corneal opacities, keratitis, band-like keratopathy, pterygia, and neovascular pannus, similar to squamous and basal cell carcinoma of the eyelid (more frequent at the margin); eyelid freckling; (de)pigmentation; scarring, leading to ectropion and loss of eyelashes; and melanomas [77, 78]. From an electrophysiologic standpoint, the flash visual evoked potential (fVEP) amplitude is markedly attenuated, while ffERG responses can range from near normal to non-recordable [79, 80].

For Cockayne's syndrome, only supportive treatments are available. A close dermatologic and ophthalmologic monitoring is indispensable (and effective) in managing the xeroderma pigmentosum-like manifestations, and protection of the skin from sunlight is indispensable to avoid the damaging effects of UV light. The prognosis, however, remains generally very poor, except for patients with deep intronic genetic variations, who appear to have a better survival expectancy [81].

Alagille Syndrome (Arteriohepatic Dysplasia)

Alagille syndrome, also known as arteriohepatic dysplasia, is an autosomal dominant (with high penetrance and variable expressivity) genetic disease first described by Alagille et al. in 1975 [82]. It is caused by mutations mainly in the *JAG1* gene [83] or, more rarely, in the *NOTCH2* gene [84]. Jagged-1, the protein derived from *JAG1,* belongs to the Notch signaling pathway playing essential roles at multiple stages of hematopoiesis and regulates also the development or the homeostasis of cells in many tissues and organs [85]. Notch-2 is part of the same signaling pathway.

The typical characteristics of Alagille syndrome are summarized in Table 7.4. The hallmark of this syndrome is a major form of childhood-onset chronic liver disease, usually from birth with neonatal jaundice [86, 87]. A combination of these liver issues with typical facies characterized by broad and prominent forehead, pointed mandible/chin, bulbous tip of the nose and deep-set eyes, skeletal anomalies, cardiovascular disorders (cardiac murmur, pulmonary artery stenosis, and other malformations), renal disease, and other less common findings characterize systemically the majority of the patients. The 20-year predicted life expectancy for these patients is 75%, and congenital heart disease is mainly associated with increased mortality as well as increased risk for intracranial bleeding and hepatic disease.

The ocular findings of Alagille syndrome include anterior and posterior segment changes (Table 7.4), and the most common changes are posterior embryotoxon, iris abnormalities, optic disc anomalies, diffuse fundus hypopigmentation, and other changes to include a frank and widespread pigmentary retinopathy [88–91]. The imaging findings of a patient with Alagille syndrome with disc and retinal findings are illustrated in Figs. 7.8 and 7.9.

From a visual electrophysiology standpoint, when a retinopathy is seen in patients with Alagille syndrome, the ffERG is markedly decreased (Fig. 7.9). The VEP response may also be abnormally reduced and/or delayed, without specific patterns of anomaly, in part due to the presence of the optic nerve head drusen [92, 93], in part potentially due to other factors such as intracranial hypertension or disc inflammation that need to be evaluated on a case-by-case basis.

Senior–Loken Syndrome

Senior–Loken syndrome (SLS) is a rare autosomal recessive condition affecting approximately 1:100,000 people, first reported independently by Senior et al. and Loken et al. in 1961 [94, 95]. This syndrome is mainly characterized by renal (medullar cystic disease called nephronophthisis that can lead to kidney failure) and retinal manifestations (in its typical description, LCA-like, but this is by no means the rule). Other non-cardinal features of SLS include cone-shaped deformity of long bone epiphyses, sensorineural hearing loss, liver fibrosis, cerebellar ataxia, and vasopressin-resistant diabetes insipidus.

At least nine genes have been implicated in causing SLS [42, 43], all of which are involved in ciliogenesis and regulation of ciliary protein trafficking, characterizing this disease as a ciliopathy [42, 96]. Unlike BBS and ALMS, though, SLS patients are not obese and do not have abnormalities of the digits. These genes interact with other ciliary genes involved in eye diseases, such as *RPGR* [96]. This finding helps explain why SLS can be associated with both RP- and CORD-type

Table 7.4 Characteristics of Alagille syndrome

Characteristic systemic features
Liver biopsy: Paucity of the interlobular ducts and fibrosis (85% of patients)[a]
Liver transplantation for hepatic decompensation necessary in 21% (19 of 92) of patients with 79% survival 1-year posttransplantation[a]
Chronic cholestasis – 96%[a]
Cardiac murmur (pulmonary valvular stenosis) – 97%[a]
Peripheral arterial stenosis also possible
Characteristic facial features (96%):[a, b]
Broad and prominent forehead
Pointed mandible/chin
Bulbous tip of the nose
Deep-set eyes
Butterfly-shaped vertebrae – 51%[a]
Renal disease – 40%[a]
Intracranial bleeding or stroke – 14%[a]
Less common features: Absent deep tendon reflexes growth and mental retardation high-pitched voice
Characteristic ocular features
Anterior segment:
Posterior embryotoxon – 78%[a] to 95%[c] of cases
Axenfeld anomaly in 13% of cases[a]
Microcornea also possible
Iris abnormalities – 45%[c]
Eccentric/ectopic pupils
Posterior segment:
Optic disc anomalies – 76%[c]
Optic nerve head drusen most frequently
Diffuse fundus hypopigmentation – 57%[a, c]
Speckled RPE appearance – 33%[a, c]
pigmentary retinopathy – 4.5%[a, c]

[a]Prevalence figures based on Emerick et al. *Hepatology* 1999 (92 cases)
[b]From Kamath et al. *Am. J. Med. Genet.* 2002
[c]Prevalence figures based on Hingorani et al. *Ophthalmology* 1999 (22 UK cases and 23 parents)

phenotypes, because RPGR is expressed in both photoreceptor types and it is well-known mutations in *RPGR* can cause both RP and CORD type of signs.

Regarding the ocular manifestations of SLS, photophobia, nystagmus, and hyperopia are common in these patients. Congenital cataracts and keratoconus can also be present and may require surgery. When observed, the LCA-like retinopathy is typically functionally severe at onset, although usually characterized by a mismatch between function and retinal microanatomical integrity, as it is certainly the case of SLS patients harboring *NHPH6* mutations, also known as *CEP290* [97–99], a gene also causing allelic forms of pure, non-syndromic LCA (LCA10) and forms

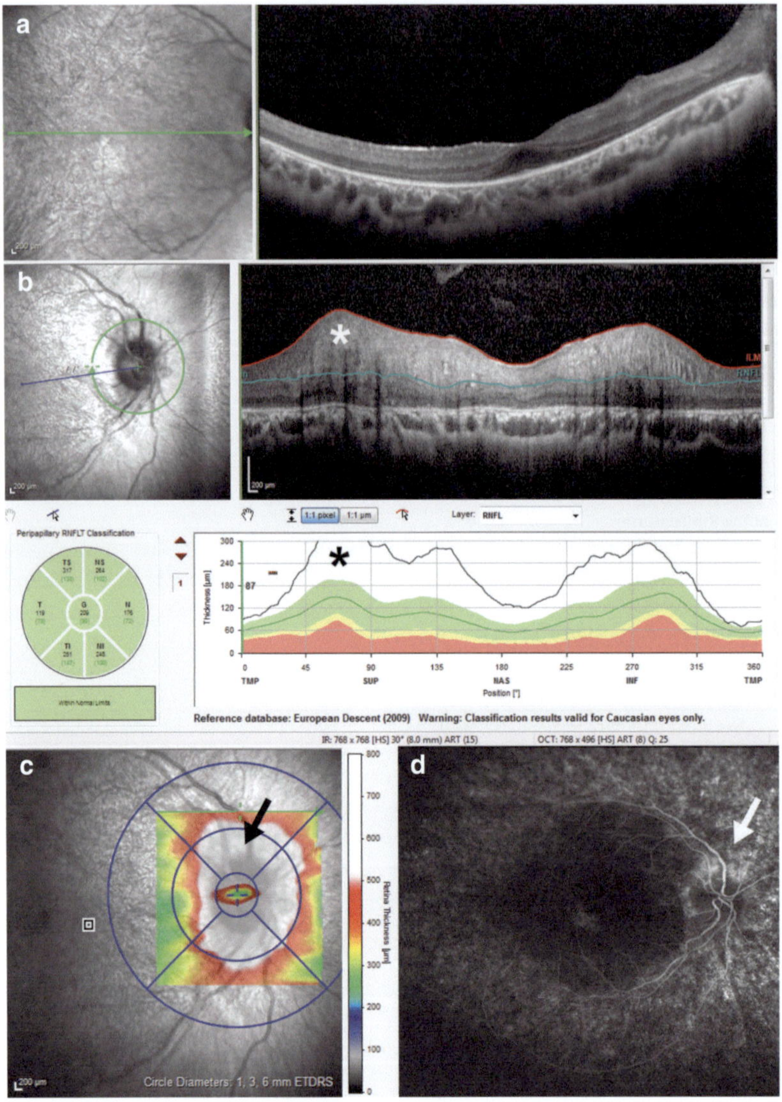

Fig. 7.8 OCT findings in Alagille syndrome. These are OCTs from a 13-year-old Hispanic female patient with night blindness and difficulty with localizing objects since she was 4 years old and a *JAG1* heterozygous disease-causing mutation. The systemic picture was remarkable for hydrocephalus at 5 years of age being treated with a shunt and liver transplant, managed with immunosuppression. (**a**) The macular OCT shows diffuse ellipoid zone (EZ) loss and outer nuclear layer (ONL) thinning but good EZ preservation in the fovea, although the intensity of the EZ signal appears very faint. (**b**) The optic nerve head OCT shows marked bilateral generalized retinal nerve fiber layer (RNFL) thickening. The area of maximal thickening is highlighted by asterisks. An enface thickness map centered on the optic nerve head is illustrated in (**c**), to show the large area of RNFL swelling. (**d**) While this patient had papilledema from the hydrocephalus, a late fluorescein angiogram also shows mild but noticeable leakage in the area of maximal RNFL swelling seen on OCT (arrows), suggesting at least a portion of the disc swelling may be attributable to optic nerve inflammation. In this patient, there was no evidence for optic nerve drusen

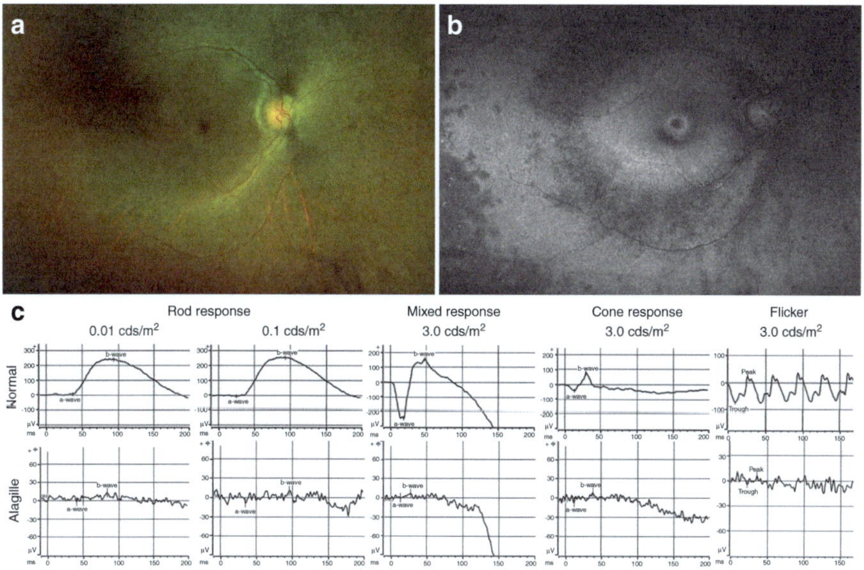

Fig. 7.9 Clinical and functional findings in Alagille syndrome. (**a**) Color fundus photography of the same patient is shown in Fig. 7.8. Peripheral disseminated nummular RPE dropout, attenuated vessels, thinned macular tissue with enhanced vitreoretinal interface reflexes, and optic nerve swelling are visible. (**b**) Fundus autofluorescence reveals multiple rings of alternating hypo- and hyper-autofluorescence with a bull's eye-like maculopathy, pericentral dystrophy at the arcades and peripheral RPE dropout with punctate hyper-AF deposits in the far temporal periphery. There is also faint hyper-autofluorescence of the optic nerve head, but no clear evidence of optic disc drusen. (**c**) The ffERG shows non-recordable rod and mixed responses and residual cone responses, consistent with an advanced rod–cone pattern of retinal degeneration of the RP family

of BBS (BBS14). A similar finding also characterizes the forms associated with *NPHP5* gene mutations [100, 101]. However, cases with fairly well-preserved visual acuity, peripheral fields, and a CORD phenotype on ffERG testing have been observed, as it is the case of the patient illustrated in Fig. 7.10, harboring a homozygous *NHNP1* whole-gene deletion.

Although SLS is rarer than LCA and BBS, due to the severity of the renal involvement, prompt recognition of this disorder has important diagnostic and prognostic implications, as kidney disease typically evolves toward failure and can require transplantation within the first few decades of life. It should be noted presymptomatic screenings for renal abnormalities can be initially normal. Thus, when the index of suspicion of SLS is high, molecular genetic diagnostic testing is essential to confirm the diagnosis. In the absence of access to genetic testing, at least yearly kidney function profile testing and ultrasounds are strongly recommended to minimize the risk of sudden-onset signs and symptoms of kidney failure. In addition, even though for now limited to a very specific truncating mutation, the recent initial evidence of possible therapeutic success with intravitreal injections of editing antisense oligonucleotides in LCA10 children makes the recognition of SLS due to *NPHP6/CEP290* mutations an even greater priority from the treatment standpoint [102].

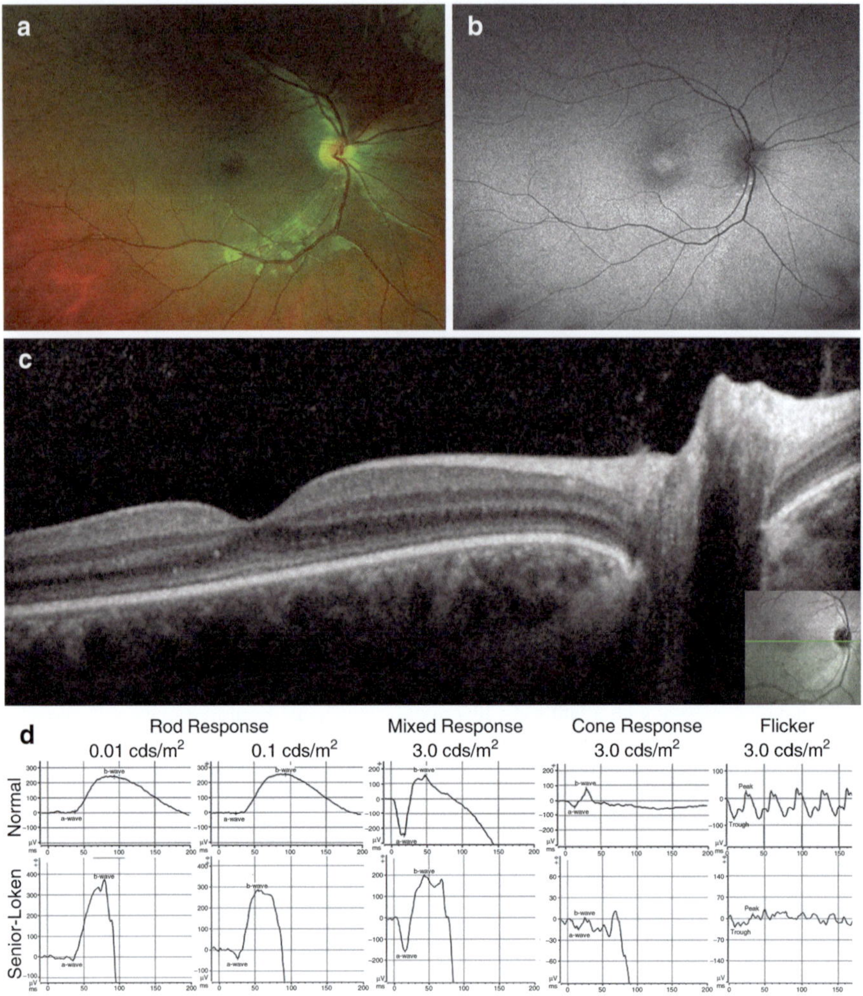

Fig. 7.10 Findings in Senior–Loken syndrome. (**a**) Fundus appearance of a 15-year-old Caucasian male with a homozygous deletion encompassing the whole *NPHP1* gene. The patient had already received a kidney transplant due to renal failure. He was referred for suspected Senior–Loken syndrome due to a complaint of photophobia, reduced visual acuity (OD 20/80, OS 20/320), and nystagmus with a strong latent component. In this young patient, fundus examination reveals no obvious pathology. (**b**) Fundus autofluorescence imaging reveals, however, a clear perifoveal hypo-autofluorescent ring and mild foveal hyper-autofluorescence. (**c**) The macular OCT shows only a loss in definition of the EZ with an otherwise normal retinal profile, lamination, and thickness and only greater than usual prominence of the interdigitation zone between the EZ and the RPE layers. (**d**) The full degree of the pathology in this patient is revealed only by ffERG testing, where there are markedly diminished cone responses with good preservation of rod responses and reduced mixed a-wave amplitude, configuring a virtually cone-only dystrophy pattern of retinal degeneration. This manifestation of Senior–Loken syndrome due to a whole *NPHP1* gene deletion is far less severe than what other patients usually exhibit (i.e., much more severe compromise of both rods and cone ffERG responses), but the mismatch between apparently well-preserved microanatomy of the retina and poor retinal function is a recurrent hallmark of this ciliopathy

Joubert Syndrome

Joubert syndrome is a rare autosomal recessive ciliopathy with an estimated incidence of 1 in 100,000, first identified in 1969 by pediatric neurologist Marie Joubert [103]. Despite its rarity, Joubert syndrome exhibits remarkable genetic heterogeneity. The genes responsible for Joubert syndrome are over 30 and largely overlap with those responsible for various forms of LCA, SLS, and BBS, with which this condition is allelic, because they are for the most part all ciliary genes. In patients with JS with ocular involvement, the genes most commonly involved are *AIH1, INPP5E, ARL13B,* and *CC2D2A* [104].

Joubert syndrome is characterized by brain abnormalities (the "molar tooth sign" on MRI imaging is pathognomonic), with retinal and renal involvement. From a neurological standpoint, this syndrome can manifest with ataxia, hyperpnea, sleep apnea, abnormal eye and tongue movements, hypotonia, and seizures. Other developmental anomalies, such as polydactyly, cleft lip or palate, and tongue abnormalities, can also occur.

As in other ophthalmic ciliopathies, Joubert syndrome has photoreceptor cells damage as the main retinal feature causing variable retinal dystrophies, mainly RP-like and LCA-like ones (Fig. 7.11), whereby the retinal phenotype can be indistinguishable from other related forms of LCA, SLS, and BBS. Other described ophthalmic manifestations include chorioretinal colobomas and a Coat's like phenotype [104–106].

Metabolic Disorders

Abetalipoproteinemia (Bassen–Kornzweig Syndrome)

Abetalipoproteinemia (ABL), also called Bassen–Kornzweig syndrome [107], is a rare autosomal recessive disease, characterized by deficiency in the absorption of fat and fat-soluble vitamins (e.g., vitamin A, D, E, and K) from food [108]. It is caused by mutations of the *MTTP* gene encoding the microsomal triglyceride transfer protein necessary for lipid metabolism [109–111].

From the systemic standpoint, ABL patients display intestinal fat malabsorption, causing with failure to thrive, diarrhea, and vomiting, progressive spinocerebellar degeneration ataxia, and typically abnormal erythrocytes on a blood smear (acanthocytosis, i.e., irregularly spiked red blood cells). At the serological level, ABL patients exhibit absent or extremely low LDL-cholesterol, triglyceride, and apolipoprotein (apo) B levels.

The ocular phenotype of ABL is characterized by a degenerative retinopathy indistinguishable from the clinical and visual electrophysiological standpoint from non-syndromic RP. In ABL patients, the rod and cone ffERG responses are markedly decreased and can become non-recordable in late disease stage. Consistent

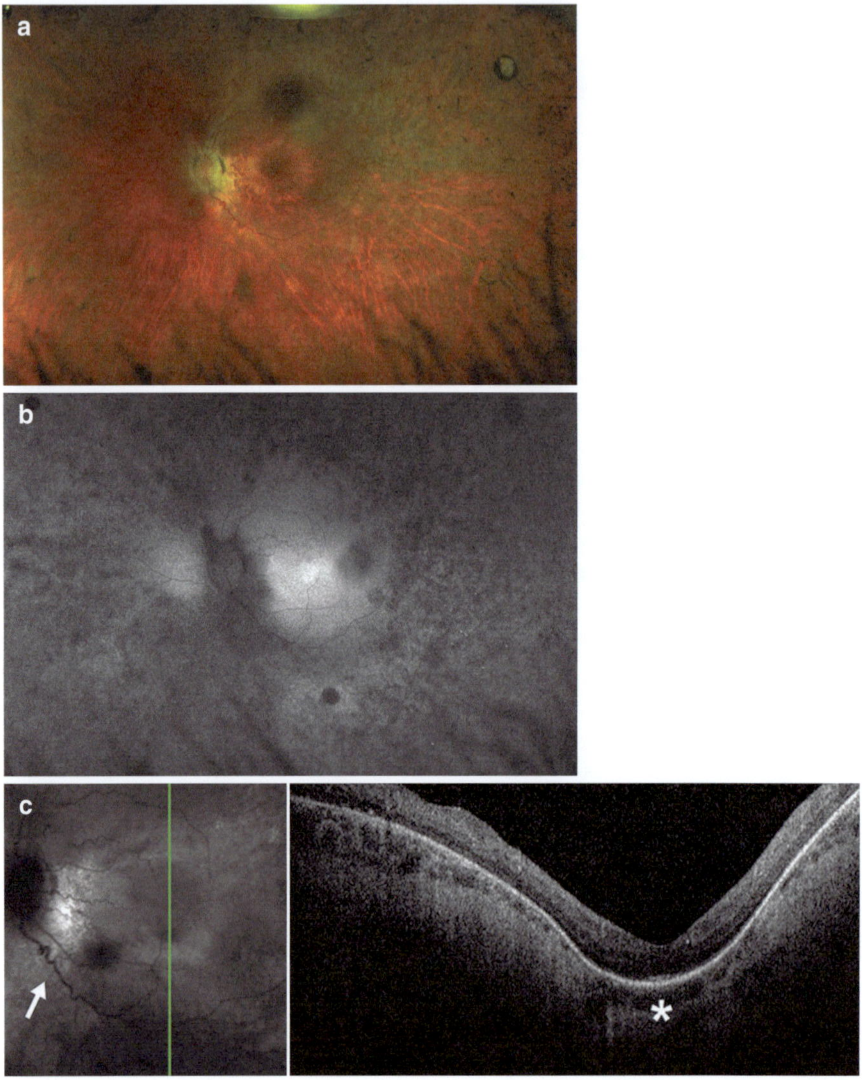

Fig. 7.11 Imaging findings in Joubert syndrome. (**a**) Fundus photograph of a 30-year-old Caucasian male patient with Joubert syndrome due to compound heterozygous mutations in the *AIH1* gene. This patient was first diagnosed with RP at 3 years of age, at the time he also developed nystagmus. His review of systems was positive for complete kidney renal failure, also developed at 3 years of age, seizures, developmental delay, and speech impediment. Fundus examination shows bone spicule-like pigment deposits, abnormally tortuous vessels (arrow), mild waxy disc pallor, and a posterior staphyloma. At this stage of the disease, acuity was 20/160 in each eye. (**b**) Fundus autofluorescence imaging shows a disseminated speckled hypo-autofluorescence at and outside of the arcades and an internal smaller patch of macular hyper-autofluorescence around macula. (**c**) A macular OCT, rendered challenging by the nystagmus, shows diffuse EZ/ELM/ONL loss, a significant posterior staphyloma, and marked choroidal thinning. The foveal region shows fair EZ preservation (asterisk)

with the significant neurodegenerative manifestations characterizing these patients, the VEP P100 latency may also be delayed [112–114].

Systemic surveillance is essential for the management of these complex yet treatable patients [111]. Management includes maintaining adequate caloric intake to alleviate growth deficiency, a low-fat diet inclusive of oral essential fatty acid supplementation, and supplementation with vitamin A, vitamin D, vitamin E, and vitamin K, preventing most of the more severe complications of ABL [111, 115]. Recognition of this unique syndrome is essential not only because it is susceptible to response to treatment but also because, unlike non-syndromic RP [116], ABL benefits exquisitely well from vitamin E supplementation and also from the retinal standpoint [115].

Neuronal Ceroid Lipofuscinoses (Batten's Disease)

Neuronal Ceroid Lipofuscinoses (NCLs), also collectively referred to as Batten's disease, are a class of genetically heterogeneous neurodegenerative disorders. The first reported description of NCL was by Otto Christian Stengel in 1826, followed by similar reports by Frederick Batten in 1903. Most forms of NCL are inherited in an autosomal recessive fashion. However, a dominant adult-onset form of NCL, termed CLN4, is caused by a distinct gene, *DNAJC5* [117, 118]. The shared pathogenetic event of all forms of NCL are genetically induced defects in lysosomal storage, leading to accumulation of an autofluorescent lipopigment – called ceroid lipofuscin – ultimately inducing neuronal degeneration of the affected tissues [119]. NCLs are generally characterized by progressive loss of vision, motor and cognitive disfunctions, seizure, and premature demise. They are currently classified by the age of disease onset as infantile NCL (INCL), late infantile NCL (LINCL), juvenile NCL (JNCL), and adult NCL (ANCL).

JNCL is the most common form of NCLs mainly associated with mutations (the most common being a 1.02 kb homozygous deletion) in the *CLN3* gene, typically showing a severe degenerative retinopathy affecting both the peripheral and the central retina [118]. This retinopathy has characteristics of both a Stargardt-like macular dystrophy and a form of RP *sine pigmento* (at least in the first two decades of life) and usually lead to a severe vision loss by 10 years of age. Two representative cases of JNCL with molecularly confirmed *CLN3* mutations are shown in Fig. 7.12. The life expectancy of JNCL patients is typically up to the late third/early fourth decade of life for patients whose disease is due to the common 1.02 kb homozygous deletion in the *CLN3* gene, whereas longer survival is expected for compound heterozygous patients. Cardiac involvement due to storage accumulation usually affecting the conduction system is relatively common and may be responsible for fatal arrhythmias. It can be prevented with a pacemaker implantation [118]. In these patients, the mfERG has also been investigated and, consistent with the severe macular involvement of JNCL, mfERG responses are markedly abnormal as well [120].

INCL is the most severe form of NCLs and usually occurs before 2 years of age with severe vision loss, psychomotor retardation, and reduced head growth that

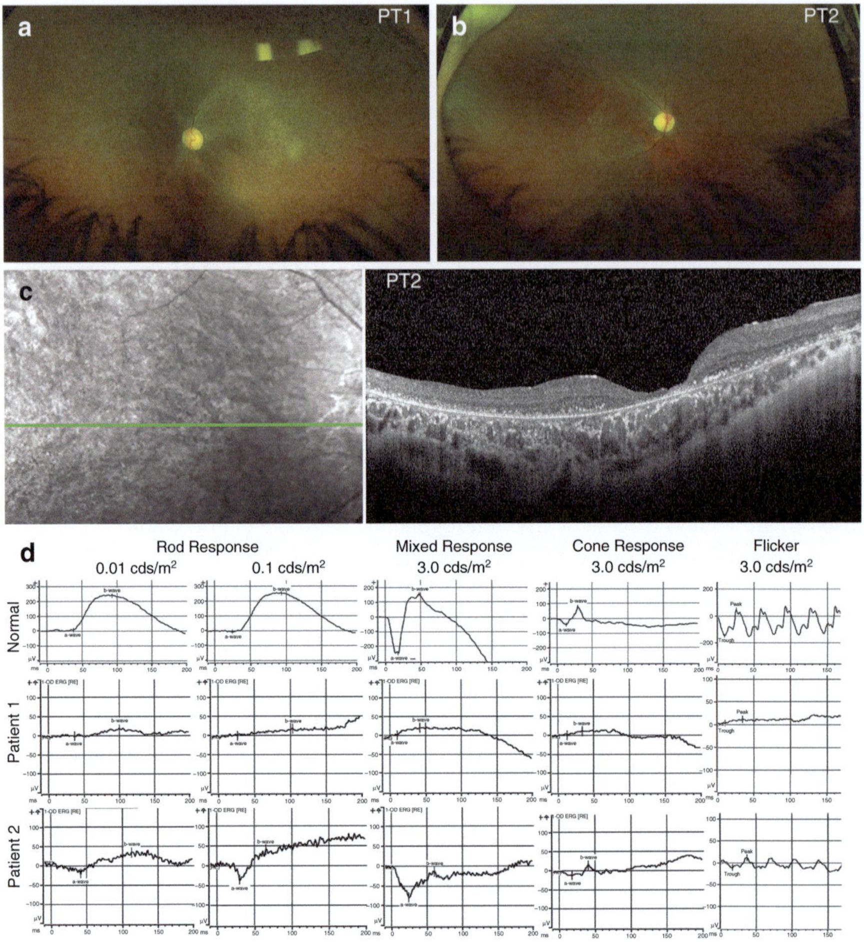

Fig. 7.12 Imaging and functional findings in Batten's disease. Images from two patients with molecularly confirmed Batten's disease (JNCL) are shown. Patient 1 (PT1) is a 12-year-old male, and patient 2 (PT2) is a 12-year-old female, both carrying *CLN3* loss-of-function mutations (homozygous 1.02 kb deletion in PT1 and compound heterozygote for the common 1.02 kb deletion and an intronic splice site mutation in PT2). Both had, in association to a history of profound and rapidly progressive vision loss, also a history of emerging neurological problems. (**a**) Color fundus photograph of PT1, remarkable for disc pallor and macular sub-atrophy with a diffuse ring of moth-eaten RPE loss with some pigment deposits across the post pole and along the arcades. (**b**) Color fundus photograph of PT2, showing beaten bronze macular appearance with para-central and nasal RPE mottling. A temporal metallic reflex is also seen. (**c**) Macular OCT of PT2 shows generalized neuroretinal thinning affecting both inner and outer retinal layers with a near bare foveal region and intraretinal hyper-reflective deposits/fragments. (**d**) The ffERG findings of both patients compared to a normal control are shown: PT1 exhibits more advanced diseases, with markedly reduced rod ERGs and non-recordable mixed or cone-driven responses, PT2 shows less severe disease, with moderately reduced rod and cone ERG responses, and electronegative mixed ERG morphology, indicative of post-receptoral dysfunction, as reported already by others (see text). In both cases, there is a pattern of cone≥rod retinal degeneration

can lead to microcephaly [121]. This form is mainly associated with the mutations in the *CLN1* gene encoding for a palmitoyl-protein thioesterase-1 (PPT1) whose loss results in severe cerebral degeneration [122]. INCL usually leads affected children to death by 10 years of age. The retinal degenerative manifestations are equally severe, but are characterized initially by electronegative ffERG due to selective b-wave truncation suggestive of post-receptoral dysfunction in the earliest disease stages, followed by greater amplitude losses in response amplitudes for both the a- and the b-wave with persistent mixed ffERG response electronegativity. These findings, alongside histological and immunohistochemical studies from the retinas of affected children who perished during the first decade of life, support the existence early in the disease process of both a pre- and/or postsynaptic neurotransmission block from photoreceptors to bipolar cells in INCL [123].

LINCL is mainly associated with mutation in the *CLN2* gene, encoding for the tripeptidyl peptidase 1 (TPP1) enzyme, and occurs between 2 and 4 years of age, when visual symptoms usually start and rapidly progress to bare light perception and severe ffERG compromise [124]. Life expectancy of untreated LINCL is usually up to no more than the early adolescence [124]. However, in 2017, cerliponase alfa (Brineura, Biomarin Pharmaceuticals), a tripeptidyl peptidase enzyme replacement therapy, became the first approved treatment for *CLN2*-linked Batten's disease. Thus, improved prognosis and prolonged survival for these patients can now be expected to occur and also new data on the longer-term effects of *CLN2* mutations on both neurological and ophthalmological phenotypes to be generated. It is not clear whether Brineura may also have an impact on the visual symptoms of LINCL patients, as the treatment is currently administered only via intraventricular infusion.

ANCL is the rarest form of NCLs, and its symptoms typically appear around 30 years of age. Consistent with the late-onset nature of ANCL and the variable and more limited disease expressivity, the life span of ANCL-affected patients can vary considerably [124]. ANCL is currently classified in two different types: Type A is associated with the mutations in the *CLN6* or *PPT1* genes and is mainly characterized by neurological symptoms and associated ocular findings. In these patients, a cone-only ffERG dystrophic pattern has been reported, with preservation of rod-driven responses, associated with a markedly abnormal EOG Arden ratio, indicative of widespread RPE dysfunction [125]. Type B ANCL is associated with mutations in the *CTSF* or *DNAJC5* genes and is usually free of visual symptoms.

In addition to the noted Brineura treatment for CLN2-associated Batten's disease, gene therapy human clinical trials are currently underway for *CLN2*-, *CLN3*-, and *CLN6*-associated Batten's disease [124].

Mucopolysaccharidosis

The mucopolysaccharidosis (MPS) syndromes are a group of lysosomal storage diseases mainly affecting the central nervous system. Various forms of enzyme replacement therapies have been recently developed for MPS and are promising to change significantly the prognosis and the clinical and functional landscape of this group of disorders [126]. There are multiple ocular manifestations associated with the various forms of MPS. Here we will focus only on the subtypes with reported retinal manifestations or with visual electrophysiological findings.

Mucopolysaccharidosis I (MPS I, also known as Hurler–Scheie syndrome), is an autosomal recessive condition caused by mutations in the *IDUA* gene encoding for alpha-L-iduronidase [127]. Signs vary including type of retinopathy: in most cases, the retinal disease is of the RP type with primary impairment of rod-driven responses and possible evidence for electronegative mixed ffERGs and relatively normal cone-driven responses. Cases with a bull's eye maculopathy and abnormalities primarily affecting the mfERG have also been reported [128, 129]. Retinopathy in MPS I has been reported in about 43% of cases and delayed VEPs in 80% [130]. However, all patients investigated in the latter study also exhibited corneal clouding that could confound VEP interpretation.

Mucopolysaccharidosis II (MPS II, also known as Hunter syndrome), is an X-linked condition due to changes in the *IDS* gene encoding for iduronate 2-sulfatase [reviewed in [131]]. A retinopathy of the RP type with electronegative ffERGs has been reported also in this subtype of MPS [132], as well as a disseminated retinal pigment epitheliopathy, found in as many as 50% of the investigated patients [130]. In the latter study, delayed VEPs potentially attributable to impaired optic pathway conduction was also found in 67% of cases and not attributable to confounding from corneal clouding as this complication was not observed in these patients.

Mucopolysaccharidosis III (MPS III), also called Sanfilippo syndrome, is an autosomal recessive lysosomal storage disease mainly affecting the central nervous system. Despite its rare occurrence, it accounts for up to 80% of all MPSs and is the most common form of MPS in the United States [133]. Sanfilippo syndrome is associated with a deficiency in lysosomal enzymes needed to break down the glycosaminoglycan heparan sulfate. MPS III is divided into four types (IIIA, IIIB, IIIC, and IIID) according to their genetic cause (*SGSH*, *NAGLU*, *HGSNAT*, or *GNS* gene, respectively). All four types have similar symptoms, except MPS IIIA which usually occurs earlier in life and progress more rapidly. MPS III typically appears during early childhood with symptoms including delayed speech and behavior problems such as anxious, aggressive, destructive, restless and sleep disturbances, progressive intellectual disability, or difficulty with social interactions and communication (autism). In later stages of MPS III, patients may also have hearing loss, progressive movement disorders, dementia, and seizures (reviewed in [133]).

Vision problems are almost constantly associated with MPS III. The most common ocular manifestations of MPS III are optic nerve swelling and optic atrophy. However, due to the deposition of heparan sulfate throughout virtually all ocular structures, clinical findings may include corneal changes (that can require transplantation), cataract formation, glaucoma, and pigmented retinopathy. In particular, in MPS IIIA, ffERG rod- and cone-driven responses decrease significantly and can have an electronegative shape and a b-wave split as shown in Fig. 7.13.

MPS IV (Morquio syndrome) is another genetically heterogeneous autosomal recessive form of MPS in which retinopathy is seldom found (6% of cases), whereas

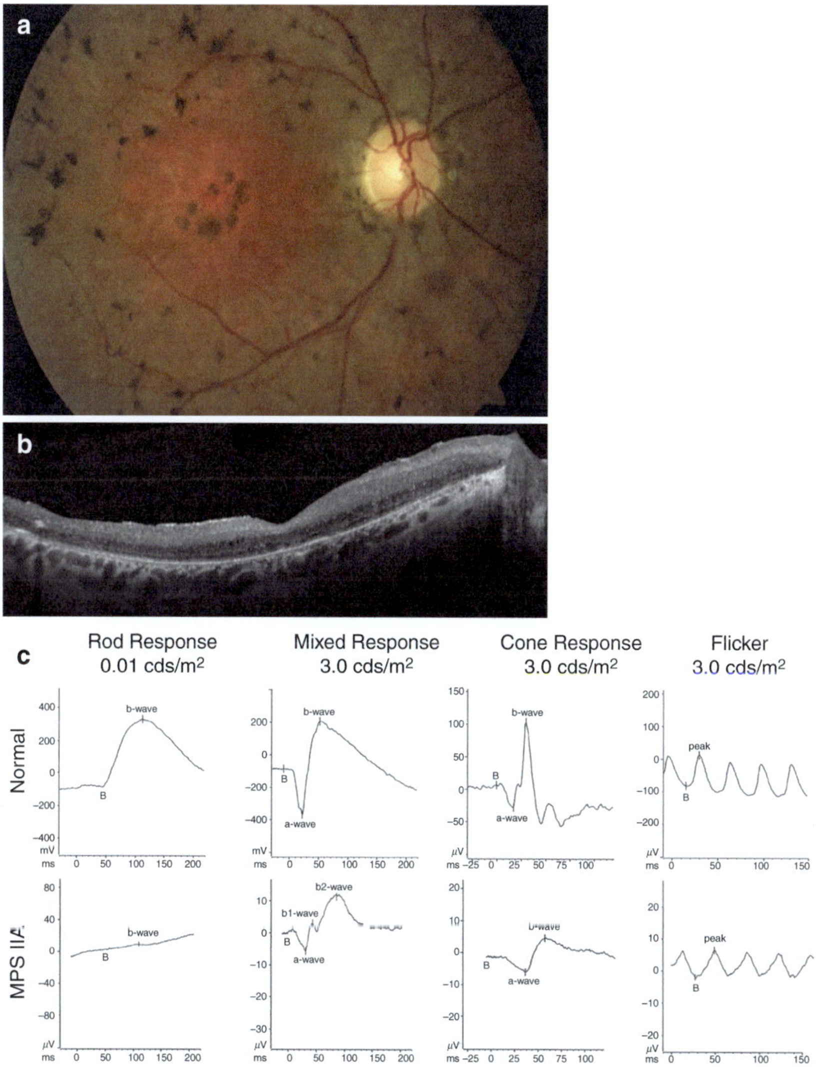

Fig. 7.13 Imaging and functional findings in Sanfilippo syndrome type IIIA. Findings in a 29-year-old Caucasian male with a 2-year history of progressive decline in vision (20/100 acuity in OD, the better eye) and night blindness, bearing a preexisting diagnosis of Sanfilippo syndrome type IIIA confirmed by molecular genetic diagnostic testing (compound heterozygote for *SGSH* gene mutations). (**a**) Fundus examination shows coarse breadcrumb-like mid-peripheral retinal pigmentary deposits, mild vascular attenuation, mild peri-papillary discoloration OU, and a ring of peri-foveal pigmentary changes. (**b**) Macular OCT scans show generalized retinal thinning; marked attenuation of the outer retinal layers with very little EZ, ELM, and ONL residual thickness; disseminated fine hyperreflective deposits; and a thickened inner retina with an amorphous appearance. Fine cystic changes (macular edema) can also be seen temporally to the fovea. (**c**) Non-recordable rod-driven ERGs markedly reduced mixed responses, with a b-wave split in a b1-like, presumably cone-driven response and a rod-driven b2-like one. Cone responses were all markedly reduced in amplitude and markedly delayed, but still clearly recordable, configuring a rod–cone RP-like pattern of retinal degeneration

delayed VEPs have been observed in 50% of cases [130]. Caution in interpreting the latter findings remains in order, however, due to the almost invariable corneal clouding.

Lastly, MPS VI (Maroteaux–Lamy syndrome) is another autosomal recessive form of MPS due to mutations in the *ARSB* gene encoding for arylsulfatase B [134]. In MPS VI, it has been reported approximately 50% of the patients exhibit a retinopathy [130]. Delayed VEPs have been reported in all cases investigated in the latter study as well. However, also these patients exhibited invariably corneal clouding as well.

References

1. Iannaccone A, Berdia J. Retinitis pigmentosa. Danbury: National Organization for Rare Disorder; 2017. Review No. 21.
2. Smith RJ, et al. Clinical diagnosis of the Usher syndromes. Usher Syndrome Consortium. Am J Med Genet. 1994;50(1):32–8.
3. Iannaccone A, et al. Clinical and immunohistochemical evidence for an X linked retinitis pigmentosa syndrome with recurrent infections and hearing loss in association with an RPGR mutation. J Med Genet. 2003;40(11):e118.
4. Iannaccone A, et al. Retinal phenotype of an X-linked pseudo-Usher syndrome in association with the G173R mutation in the RPGR gene. Adv Exp Med Biol. 2008;613:221–7.
5. Edwards A, et al. Visual acuity and visual field impairment in Usher syndrome. Arch Ophthalmol. 1998;116(2):165–8.
6. Fishman GA, et al. Prevalence of foveal lesions in type 1 and type 2 Usher's syndrome. Arch Ophthalmol. 1995;113:770–3.
7. Iannaccone A. Usher syndrome: correlation between visual field size and maximal ERG response b-wave amplitude. In: LaVail MM, Hollyfield JG, Anderson RE, editors. Retinal degenerations: mechanisms and experimental therapy. New York: Plenum Publishers; 2003. p. 123–31.
8. Iannaccone A, et al. Kinetics of visual field loss in Usher syndrome Type II. Invest Ophthalmol Vis Sci. 2004;45(3):784–92.
9. Herrera W, et al. Retinal disease in Usher syndrome III caused by mutations in the clarin-1 gene. Invest Ophthalmol Vis Sci. 2008;49(6):2651–60.
10. Refsum S, Salomonsen L, Skatvedt M. Heredopathia atactica polyneuritiformis in children. J Pediatr. 1949;35:335–43.
11. Wanders RJA, Waterham HR, Leroy BP. Refsum disease. In: Adam MP, et al., editors. GeneReviews(R). University of Washington, Seattle; 1993.
12. Gibberd FB, et al. Heredopathia atactica polyneuritiformis (refsum's disease) treated by diet and plasma-exchange. Lancet. 1979;1(8116):575–8.
13. Strom TM, et al. Diabetes insipidus, diabetes mellitus, optic atrophy and deafness (DIDMOAD) caused by mutations in a novel gene (wolframin) coding for a predicted transmembrane protein. Hum Mol Genet. 1998;7(13):2021–8.
14. Inoue H, et al. A gene encoding a transmembrane protein is mutated in patients with diabetes mellitus and optic atrophy (Wolfram syndrome). Nat Genet. 1998;20:143–8.
15. Mozzillo E, et al. A novel CISD2 intragenic deletion, optic neuropathy and platelet aggregation defect in Wolfram syndrome type 2. BMC Med Genet. 2014;15:88.
16. Amr S, et al. A homozygous mutation in a novel zinc-finger protein, ERIS, is responsible for Wolfram syndrome 2. Am J Hum Genet. 2007;81(4):673–83.

17. Lucariello A, et al. Modulation of wolframin expression in human placenta during pregnancy: comparison among physiological and pathological states. Biomed Res Int. 2014;2014:985478.
18. De Falco M, et al. Localization and distribution of wolframin in human tissues. Front Biosci (Elite Ed). 2012;4:1986–98.
19. Borgna-Pignatti C, et al. Thiamine-responsive anemia in DIDMOAD syndrome. J Pediatr. 1989;114(3):405–10.
20. Kozak I, et al. New observations regarding the retinopathy of genetically confirmed Kearns-Sayre syndrome. Retin Cases Brief Rep. 2018 Fall;12(4):349–58.
21. Proto F, et al. Kearns-Sayre syndrome (KSS): case report and review of the literature [Italian]. Ann Ottalmol Clin Ocul. 1994;120:149–54.
22. Ohkoshi K, et al. Corneal endothelium in a case of mitochondrial encephalomyopathy (Kearns-Sayre syndrome). Cornea. 1989;8(3):210–4.
23. Finsterer J, Zarrouk-Mahjoub S. Corneal involvement in Kearns-Sayre syndrome responsive to coenzyme-Q? Cornea. 2016;35(12):e39.
24. Kim J, et al. Coenzyme Q10 in the treatment of corneal edema in Kearns-Sayre: is there an application in fuchs endothelial corneal dystrophy? Cornea. 2016;35(9):1250–4.
25. Ota I, Miyake Y, Awaya S. Studies of ocular fundus and visual functions in Kearns-Sayre syndrome – with special reference to the new stage classification. Nippon Ganka Gakkai Zasshi. 1989;93(3):329–38.
26. Holt IJ, et al. A new mitochondrial disease associated with mitochondrial DNA heteroplasmy. Am J Hum Genet. 1990;46(3):428–33.
27. Chowers I, et al. Cone and rod dysfunction in the NARP syndrome. Br J Ophthalmol. 1999;83(2):190–3.
28. Fryer A, et al. Mitochondrial DNA 8993 (NARP) mutation presenting with a heterogeneous phenotype including 'cerebral palsy'. Arch Dis Child. 1994;71(5):419–22.
29. Lemoine S, et al. Renal involvement in neuropathy, ataxia, retinitis pigmentosa (NARP) syndrome: a case report. Am J Kidney Dis. 2018;71(5):754–7.
30. Reardon W, et al. Diabetes mellitus associated with a pathogenic point mutation in mitochondrial DNA. Lancet. 1992;340(8832):1376–9.
31. Ballinger SW, et al. Maternally transmitted diabetes and deafness associated with a 10.4 kb mitochondrial DNA deletion. Nat Genet. 1992;1(1):11–5.
32. Guillausseau PJ, et al. Maternally inherited diabetes and deafness: a multicenter study. Ann Intern Med. 2001;134(9 Pt 1):721–8.
33. Massin P, et al. Prevalence of macular pattern dystrophy in maternally inherited diabetes and deafness. GEDIAM Group. Ophthalmology. 1999;106(9):1821–7.
34. Ogun O, Sheldon C, Barton JJ. Pearls & oy-sters: maternally inherited diabetes and deafness presenting with ptosis and macular pattern dystrophy. Neurology. 2012;79(6):e54–6.
35. Feigl B, Morris CP. Visual function and risk genotypes in maternally inherited diabetes and deafness. Can J Ophthalmol. 2013;48(5):e111–4.
36. Rath PP, et al. Characterisation of the macular dystrophy in patients with the A3243G mitochondrial DNA point mutation with fundus autofluorescence. Br J Ophthalmol. 2008;92(5):623–9.
37. Strauss DS, Freund KB. Diagnosis of maternally inherited diabetes and deafness (mitochondrial A3243G mutation) based on funduscopic appearance in an asymptomatic patient. Br J Ophthalmol. 2012;96(4):604.
38. Bellmann C, et al. Localized retinal electrophysiological and fundus autofluorescence imaging abnormalities in maternal inherited diabetes and deafness. Invest Ophthalmol Vis Sci. 2004;45(7):2355–60.
39. Iannaccone A. The genetics of hereditary retinopathies and optic neuropathies. Comp Ophthalmol Update. 2005;5:39–62.
40. Iannaccone A, et al. Clinical evidence of decreased olfaction in Bardet-Biedl syndrome caused by a deletion in the BBS4 gene. Am J Med Genet A. 2005;132(4):343–6.
41. Barnett S, et al. Behavioural phenotype of Bardet-Biedl syndrome. J Med Genet. 2002;39(12):e76.

42. Adams NA, Awadein A, Toma HS. The retinal ciliopathies. Ophthalmic Genet. 2007;28(3):113–25.
43. RetNet – Retinal Information Network., https://sph.uth.edu/retnet/home.htm. 2019.
44. Iannaccone A, et al. Electroretinographic alterations in the Laurence-Moon-Bardet-Biedl phenotype. Acta Ophthalmol Scand. 1996;74:8–13.
45. Iannaccone A, et al. The ocular phenotype of the Bardet-Biedl syndrome. Comparison to non-syndromic retinitis pigmentosa. Ophthalmic Genet. 1997;18:13–26.
46. Cox KF, et al. Phenotypic expression of Bardet-Biedl syndrome in patients homozygous for the common M390R mutation in the BBS1 gene. Vision Res. 2012;75:77–87.
47. Praidou A, et al. Multifocal electroretinogram contributes to differentiation of various clinical pictures within a family with Bardet-Biedl syndrome. Eye (Lond). 2014;28(9):1136–42.
48. Forte R, et al. The optic nerve in patients with the Laurence-Moon-Bardet-Biedl phenotype: a clinical and functional study. [Italian]. Boll Ocul. 1996;75(Suppl 4):115–24.
49. Marshall JD, et al. Spectrum of ALMS1 variants and evaluation of genotype-phenotype correlations in Alstrom syndrome. Hum Mutat. 2007;28(11):1114–23.
50. Marshall JD, et al. Alstrom syndrome: genetics and clinical overview. Curr Genomics. 2011;12(3):225–35.
51. Marshall JD, et al. Alstrom syndrome. Eur J Hum Genet. 2007;15(12):1193–202.
52. Hearn T. ALMS1 and Alstrom syndrome: a recessive form of metabolic, neurosensory and cardiac deficits. J Mol Med (Berl). 2019;97(1):1–17.
53. Collin GB, et al. Alms1-disrupted mice recapitulate human Alstrom syndrome. Hum Mol Genet. 2005;14(16):2323–33.
54. Katagiri S, et al. Whole-exome sequencing identifies a novel ALMS1 mutation (p.Q2051X) in two Japanese brothers with Alstrom syndrome. Mol Vis. 2013;19:2393–406.
55. Karska-Basta I, et al. Alstrom syndrome – a case report and literature review. Klin Ocz. 2008;110(4–6):188–92.
56. Malm E, et al. Full-field electroretinography and marked variability in clinical phenotype of Alstrom syndrome. Arch Ophthalmol. 2008;126(1):51–7.
57. Sadowski B, et al. Onset of bilateral blindness in the first year of life. Alstrom syndrome. Ophthalmologe. 2004;101(3):298–300.
58. Hung YJ, et al. Alstrom syndrome in two siblings. J Formos Med Assoc. 2001;100(1):45–9.
59. Russell-Eggitt IM, et al. Alstrom syndrome. Report of 22 cases and literature review. Ophthalmology. 1998;105(7):1274–80.
60. Michaud JL, et al. Natural history of Alstrom syndrome in early childhood: onset with dilated cardiomyopathy. J Pediatr. 1996;128(2):225–9.
61. Tremblay F, et al. Longitudinal study of the early electroretinographic changes in Alstrom's syndrome. Am J Ophthalmol. 1993;115(5):657–65.
62. Sheck L, et al. Alstrom syndrome – an uncommon cause of early childhood retinal dystrophy. BMJ Case Rep. 2011;2011.
63. Chandler KE, et al. Diagnostic criteria, clinical characteristics, and natural history of Cohen syndrome. J Med Genet. 2003;40(4):1–3.
64. Kolehmainen J, et al. Cohen syndrome is caused by mutations in a novel gene, COH1, encoding a transmembrane protein with a presumed role in vesicle-mediated sorting and intracellular protein transport. Am J Hum Genet. 2003;72(6):1359–69.
65. Wang H, et al. Cohen syndrome. In: Pagon RA, et al., editors. GeneReviews(R). University of Washington, Seattle; 1993.
66. Dastan J, et al. Exome sequencing identifies pathogenic variants of VPS13B in a patient with familial 16p11.2 duplication. BMC Med Genet. 2016;17(1):78.
67. Gillentine MA, Schaaf CP, Patel A. The importance of phase analysis in multiexon copy number variation detected by aCGH in autosomal recessive disorder loci. Am J Med Genet A. 2017;173(9):2485–8.
68. Rodrigues JM, et al. Cohen syndrome: review of the literature. Cureus. 2018;10(9):e3330.

69. Parri V, et al. High frequency of COH1 intragenic deletions and duplications detected by MLPA in patients with Cohen syndrome. Eur J Hum Genet. 2010;18(10):1133–40.
70. Yang C, et al. Gene analysis: a rare gene disease of intellectual deficiency-Cohen syndrome. Int J Dev Neurosci. 2018;68:83–8.
71. Chandler KE, et al. The ophthalmic findings in Cohen syndrome. Br J Ophthalmol. 2002;86(12):1395–8.
72. Hennies HC, et al. Allelic heterogeneity in the COH1 gene explains clinical variability in Cohen syndrome. Am J Hum Genet. 2004;75(1):138–45.
73. Hanawalt PC, DNA r. The bases for Cockayne syndrome. Nature. 2000;405(6785):415–6.
74. Friedberg EC. Cockayne syndrome – a primary defect in DNA repair, transcription, both or neither? BioEssays. 1996;18(9):731–8.
75. Venema J, et al. The genetic defect in Cockayne syndrome is associated with a defect in repair of UV-induced DNA damage in transcriptionally active DNA. Proc Natl Acad Sci U S A. 1990;87(12):4707–11.
76. Stefanini M, et al. Genetic analysis of twenty-two patients with Cockayne syndrome. Hum Genet. 1996;97(4):418–23.
77. Ozdirim E, et al. Cockayne syndrome: review of 25 cases. Pediatr Neurol. 1996;15(4):312–6.
78. Dollfus H, et al. Ocular manifestations in the inherited DNA repair disorders. Surv Ophthalmol. 2003;48(1):107–22.
79. Scaioli V, D'Arrigo S, Pantaleoni C. Unusual neurophysiological features in Cockayne's syndrome: a report of two cases as a contribution to diagnosis and classification. Brain and Development. 2004;26(4):273–80.
80. Ikeda N, et al. Nondetectable cone and rod electroretinographic responses in a patient with Cockayne syndrome. Jpn J Ophthalmol. 1995;39(4):420–3.
81. Schalk A, et al. Deep intronic variation in splicing regulatory element of the ERCC8 gene associated with severe but long-term survival Cockayne syndrome. Eur J Hum Genet. 2018;26(4):527–36.
82. Alagille D, et al. Hepatic ductular hypoplasia associated with characteristic facies, vertebral malformations, retarded physical, mental, and sexual development, and cardiac murmur. J Pediatr. 1975;86(1):63–71.
83. Oda T, et al. Mutations in the human Jagged1 gene are responsible for Alagille syndrome. Nat Genet. 1997;16(3):235–42.
84. Kamath BM, et al. NOTCH2 mutations in Alagille syndrome. J Med Genet. 2012;49(2):138–44.
85. Grochowski CM, Loomes KM, Spinner NB. Jagged1 (JAG1): structure, expression, and disease associations. Gene. 2016;576(1 Pt 3):381–4.
86. Emerick KM, et al. Features of Alagille syndrome in 92 patients: frequency and relation to prognosis. Hepatology. 1999;29(3):822–9.
87. Kamath BM, et al. Facial features in Alagille syndrome: specific or cholestasis facies? Am J Med Genet. 2002;112(2):163–70.
88. Puklin JE, et al. Anterior segment and retinal pigmentary abnormalities in arteriohepatic dysplasia. Ophthalmology. 1981;88(4):337–47.
89. Romanchuk KG, Judisch GF, LaBrecque DR. Ocular findings in arteriohepatic dysplasia (Alagille's syndrome). Can J Ophthalmol. 1981;16(2):94–9.
90. Mayer U, Grosse KP. [Clinical picture and inheritance of ocular symptoms in arteriohepatic dysplasia (author's transl)]. Klin Monatsbl Augenheilkd. 1982;180(4):290–3.
91. Hingorani M, et al. Ocular abnormalities in Alagille syndrome. Ophthalmology. 1999;106(2):330–7.
92. Tanino T, et al. Electrophysiological findings in a family with congenital arteriohepatic dysplasia (Alagille syndrome). Doc Ophthalmol. 1986;63(1):83–9.
93. Alvarez F, et al. Nervous and ocular disorders in children with cholestasis and vitamin A and E deficiencies. Hepatology. 1983;3(3):410–4.
94. Senior B, Friedmann AI, Braudo JL. Juvenile familial nephropathy with tapetoretinal degeneration. A new oculorenal dystrophy. Am J Ophthalmol. 1961;52:625–33.

95. Loken AC, et al. Hereditary renal dysplasia and blindness. Acta Paediatr. 1961;50:177–84.
96. Ronquillo CC, Bernstein PS, Baehr W. Senior-Loken syndrome: a syndromic form of retinal dystrophy associated with nephronophthisis. Vision Res. 2012;75:88–97.
97. Shimada H, et al. In vitro modeling using ciliopathy-patient-derived cells reveals distinct cilia dysfunctions caused by CEP290 mutations. Cell Rep. 2017;20(2):384–96.
98. Boye SE, et al. Natural history of cone disease in the murine model of Leber congenital amaurosis due to CEP290 mutation: determining the timing and expectation of therapy. PLoS One. 2014;9(3):e92928.
99. Jacobson SG, et al. Outcome measures for clinical trials of Leber congenital amaurosis caused by the intronic mutation in the CEP290 gene. Invest Ophthalmol Vis Sci. 2017;58(5):2609–22.
100. Cideciyan AV, et al. Cone photoreceptors are the main targets for gene therapy of NPHP5 (IQCB1) or NPHP6 (CEP290) blindness: generation of an all-cone Nphp6 hypomorph mouse that mimics the human retinal ciliopathy. Hum Mol Genet. 2011;20(7):1411–23.
101. Stone EM, et al. Variations in NPHP5 in patients with nonsyndromic leber congenital amaurosis and Senior-Loken syndrome. Arch Ophthalmol. 2011;129(1):81–7.
102. Cideciyan AV, et al. Effect of an intravitreal antisense oligonucleotide on vision in Leber congenital amaurosis due to a photoreceptor cilium defect. Nat Med. 2019;25(2):225–8.
103. Joubert M, et al. Familial agenesis of the cerebellar vermis. A syndrome of episodic hyperpnea, abnormal eye movements, ataxia, and retardation. Neurology. 1969;19(9):813–25.
104. Wang SF, et al. Review of ocular manifestations of Joubert syndrome. Genes (Basel). 2018;9:12.
105. Brancati F, Dallapiccola B, Valente EM. Joubert syndrome and related disorders. Orphanet J Rare Dis. 2010;5:20.
106. Parisi MA. Clinical and molecular features of Joubert syndrome and related disorders. Am J Med Genet C Semin Med Genet. 2009;151C(4):326–40.
107. Bassen FA, Kornzweig AL. Malformation of the erythrocytes in a case of atypical retinitis pigmentosa. Blood. 1950;5(4):381–7.
108. Benayoun L, et al. Abetalipoproteinemia in Israel: evidence for a founder mutation in the Ashkenazi Jewish population and a contiguous gene deletion in an Arab patient. Mol Genet Metab. 2007;90(4):453–7.
109. Shoulders CC, et al. Abetalipoproteinemia is caused by defects of the gene encoding the 97 kDa subunit of a microsomal triglyceride transfer protein. Hum Mol Genet. 1993;2(12):2109–16.
110. Wang J, Hegele RA. Microsomal triglyceride transfer protein (MTP) gene mutations in Canadian subjects with abetalipoproteinemia. Hum Mutat. 2000;15(3):294–5.
111. Burnett JR, Hooper AJ, Hegele RA. Abetalipoproteinemia. In: Adam MP, et al., editors. GeneReviews(R). Seattle: University of Washington; 1993.
112. Harcourt B, Hopkins D. Tapetoretinal degeneration in childhood presenting as a disturbance of behaviour. Br Med J. 1972;1(5794):202–5.
113. Bishara S, et al. Combined vitamin A and E therapy prevents retinal electrophysiological deterioration in abetalipoproteinaemia. Br J Ophthalmol. 1982;66(12):767–70.
114. Brin MF, et al. Electrophysiologic features of abetalipoproteinemia: functional consequences of vitamin E deficiency. Neurology. 1986;36(5):669–73.
115. Grant CA, Berson EL. Treatable forms of retinitis pigmentosa associated with systemic neurological disorders. Int Ophthalmol Clin. 2001;41(1):103–10.
116. Berson EL, et al. A randomized trial of vitamin A and vitamin E supplementation for retinitis pigmentosa. Arch Ophthalmol. 1993;111:761–72.
117. Bennett MJ, Hofmann SL. The neuronal ceroid-lipofuscinoses (Batten disease): a new class of lysosomal storage diseases. J Inherit Metab Dis. 1999;22(4):535–44.
118. Anderson GW, Goebel HH, Simonati A. Human pathology in NCL. Biochim Biophys Acta. 2013;1832(11):1807–26.
119. Dolisca SB, et al. Batten disease: clinical aspects, molecular mechanisms, translational science, and future directions. J Child Neurol. 2013;28(9):1074–100.

120. Preising MN, et al. Ocular morphology and function in juvenile neuronal ceroid lipofuscinosis (CLN3) in the first decade of life. Ophthalmic Genet. 2017;38(3):252–9.
121. Santavuori P, et al. Infantile neuronal ceroid-lipofuscinosis (INCL): diagnostic criteria. J Inherit Metab Dis. 1993;16(2):227–9.
122. Lu JY, Verkruyse LA, Hofmann SL. Lipid thioesters derived from acylated proteins accumulate in infantile neuronal ceroid lipofuscinosis: correction of the defect in lymphoblasts by recombinant palmitoyl-protein thioesterase. Proc Natl Acad Sci U S A. 1996;93(19):10046–50.
123. Weleber RG, et al. Electroretinographic and clinicopathologic correlations of retinal dysfunction in infantile neuronal ceroid lipofuscinosis (infantile Batten disease). Mol Genet Metab. 2004;83(1–2):128–37.
124. Johnson TB, et al. Therapeutic landscape for Batten disease: current treatments and future prospects. Nat Rev Neurol. 2019;15(3):161–78.
125. Dawson WW, et al. Disease-specific electrophysiological findings in adult ceroid-lipofuscinosis (Kufs disease). Doc Ophthalmol. 1985;60(2):163–71.
126. Safary A, et al. Targeted enzyme delivery systems in lysosomal disorders: an innovative form of therapy for mucopolysaccharidosis. Cell Mol Life Sci. 2019;76:3363.
127. Scott HS, et al. A common mutation for mucopolysaccharidosis type I associated with a severe Hurler syndrome phenotype. Hum Mutat. 1992;1(2):103–8.
128. Sornalingam K, et al. Variability in the ocular phenotype in mucopolysaccharidosis. Br J Ophthalmol. 2019;103(4):504–10.
129. Mack HG, Symons RCA, de Jong G. Bull's eye maculopathy and subfoveal deposition in two mucopolysaccharidosis type I patients on long-term enzyme replacement therapy. Am J Ophthalmol Case Rep. 2018;9:1–6.
130. Lin HY, et al. Ophthalmologic manifestations in Taiwanese patients with mucopolysaccharidoses. Mol Genet Genomic Med. 2019;7(5):e00617.
131. Muenzer J, et al. Multidisciplinary management of Hunter syndrome. Pediatrics. 2009;124(6):e1228–39.
132. Salvucci IDM, et al. Multimodal image analysis of the retina in Hunter syndrome (mucopolysaccharidosis type II): case report. Ophthalmic Genet. 2018;39(1):103–7.
133. Wilkin J, et al. Characterization of a case of pigmentary retinopathy in Sanfilippo syndrome type IIIA associated with compound heterozygous mutations in the SGSH gene. Ophthalmic Genet. 2016;37(2):217–27.
134. Wicker G, et al. Mucopolysaccharidosis VI (Maroteaux-Lamy syndrome). An intermediate clinical phenotype caused by substitution of valine for glycine at position 137 of arylsulfatase B. J Biol Chem. 1991;266(32):21386–91.

Chapter 8
Characteristics of Visual Electrophysiology in Inflammatory Disorders

Alessandro Iannaccone, Alfonso Senatore, Wajiha Jurdi Kheir, Donnell Creel, and Minzhong Yu

Acute Zonal Occult Outer Retinopathy (AZOOR)

Acute zonal occult outer retinopathy (AZOOR) was first reported by Gass [1]. It is an acquired inflammatory condition with symptoms of acute and rapid loss of one or more large zones of outer retinal function, photopsia, central scotomas, and normal to minimally abnormal funduscopic appearance.

Significant progress has been made in recent years in identifying the imaging characteristics of AZOOR, especially since the advent of fundus autofluorescence (FAF) [2–6]. Dense areas of markedly abnormal FAF findings, especially around the optic nerve head and along one or more of the vascular arcades, have been uniformly documented in AZOOR patients even when the ophthalmoscopic findings are absent or minimal (Fig. 8.1).

The functional electroretinographic characteristics of AZOOR were initially defined by Jacobson et al. [7], highlighting the presence of interocular asymmetry in the ffERG compromise as a hallmark and, importantly, evidence of spontaneous recovery or disease stabilization during the follow-up period in several patients. Psychophysical testing on affected patients also revealed the patchy nature of the outer retinopathy that AZOOR patients experience. These disease characteristics have been expanded and refined further, to include evidence that AZOOR patients also have abnormal EOG light peaks and Arden ratios, revealing the presence not only of an outer retinopathy but also a pigment epitheliopathy in AZOOR [8].

A. Iannaccone · A. Senatore · W. J. Kheir
Center for Retinal Degenerations and Ophthalmic Genetic Diseases, Duke University School of Medicine, Duke Eye Center, Department of Ophthalmology, Durham, NC, USA

D. Creel
Moran Eye Center, University of Utah School of Medicine, Salt Lake City, UT, USA

M. Yu (✉)
Department of Ophthalmology, University Hospitals Eye Institute, Cleveland, OH, USA
e-mail: minzhong.yu@uhhospitals.org

M. Yu et al. (eds.), *Handbook of Clinical Electrophysiology of Vision*,
https://doi.org/10.1007/978-3-030-30417-1_8

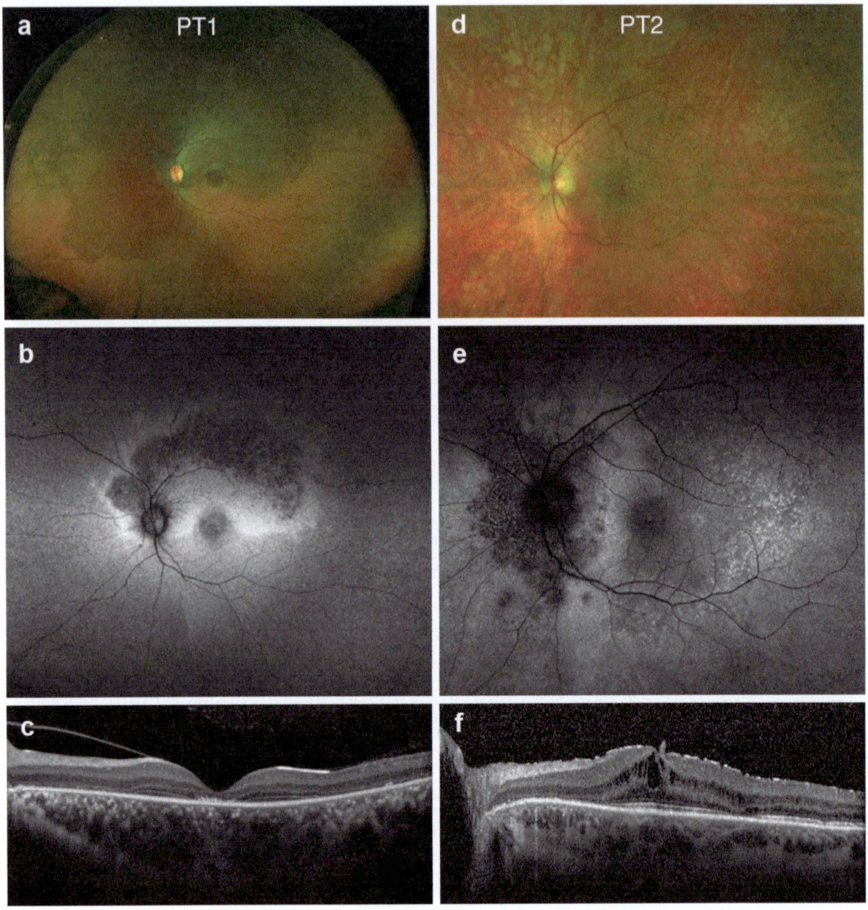

Fig. 8.1 Clinical and imaging findings in acute zonal occult outer retinopathy (AZOOR). Patient (PT) 1 is a 17-year-old female patient and PT2 a 36-year-old male patient, both with acute onset of monocular left-sided vision loss, central/paracentral scotomas, and photopsia. (**a**) Fundus findings in PT1 show a large area of discoloration along the superior arcade and around the disc. (**b**) Fundus autofluorescence (FAF) imaging shows a large area of speckled hypo-autofluorescence along the supero-temporal arcade with a wide hyper-autofluorescent halo enveloping the disc and wrapping around the fovea, with faint foveal hyper-autofluorescence. (**c**) A horizontal OCT scan shows marked ellipsoid zone (EZ), ELM, and ONL loss nasal and temporal to the fovea, central EZ crumbling, but good preservation of all layers right around the fovea. A vertical scan (not shown) found also significant EZ, ELM, and ONL loss superiorly but none inferiorly. (**d**) Fundus findings in PT2 show only faint peripapillary and superior patchy depigmentation (note how the superior hemiretina appears hypopigmented compared with the inferior one). (**e**) FAF imaging was much more informative and conclusive, revealing a large patch of speckled hypo-autofluorescence around the disc and a fainter one centrally, surrounded by a halo of speckled/punctate hyper-autofluorescence. (**f**) The OCT scan is remarkable not only for focal areas of EZ, ELM, and ONL loss that again correspond well to the area of hypo-autofluorescence nasal to the fovea and for EZ fragmentation temporal to the fovea but also for intraretinal hyporeflective cystic changes consistent with cystoid macular edema (verified also by fluorescein angiography, not shown) (Images courtesy of Dr. Dilraj Grewal, MD, Duke Eye Center)

Evidence for attenuated and delayed flicker ERGs and abnormal ON- and OFF-responses has also been identified [8], thereby pointing to a more global retinal involvement in AZOOR than purely outer retinal one (Fig. 8.2). The multifocal electroretinography (mfERG) is also informative in this condition, since patches of decreased amplitude and delayed mfERG response tend to correspond well with the areas of central visual field loss [6, 9–13] (Fig. 8.3). Accordingly, also the pattern ERG, a sensitive indicator of macular health, is usually attenuated in AZOOR [8].

There is significant clinical and symptomatic overlap between AZOOR, multiple evanescent white dot syndrome (MEWDS), ocular histoplasmosis syndrome (OHS), multifocal choroiditis and panuveitis, acute macular neuroretinopathy (AMN), acute idiopathic blind spot enlargement syndrome (AIBSES), punctate inner choroiditis (PIC), and autoimmune retinopathy (AIR), to the point that several authors have raised the possibility these conditions may represent different facets of the same disease spectrum, and many refer to AZOOR as the "AZOOR complex", reflecting well the great of overlap between AZOOR and these other conditions [8, 14–17]. To add to the degree of overlap between these entities, and especially with AIR, in recent years evidence has been provided that AZOOR patients exhibit antiretinal and anti-RPE autoantibodies (AAbs) [13, 18, 19]. Furthermore, a history of autoimmune disease has been ascertained in nearly one third of AZOOR patients [20, 21]. While symptoms may slightly differ, there may be specific diagnostic criteria to point more in the AZOOR direction than in that of AIR, and AZOOR may have a greater inclination to being a self-delimiting condition than AIR; these conditions may share an autoimmune pathogenesis, similar to other related and better established conditions such as birdshot chorioretinopathy [17, 22–25] and the paraneoplastic syndromes, cancer-associated retinopathy (CAR), melanoma-associated retinopathy (MAR) [17, 26–30], and the recently identified basal cell carcinoma-associated retinopathy and optic neuropathy (BARN) complex which, in its full expression, includes also a pigment epitheliopathy (BARRN) [31]. There is still no consensus to this date as to these conditions should be all lumped under the grand umbrella of the autoimmune retinopathies or whether splitting them each into independent categories as they were initially identified remains the most appropriate approach.

Accordingly, there is currently no clear management and treatment consensus for AZOOR. Initially, the condition was not treated and purely observed. A long-term follow-up study by Gass et al. showed that the majority of patients, while often affected initially in unilateral fashion, experience involvement of the fellow eye and recurrences [20]. In addition, while the disease did not exhibit a major tendency to progress over time after the initial insult, with stabilization of vision and visual field loss in about three-fourth of the patients, all investigated patients experienced permanent sequelae from the disease, and spontaneous improvement was seen in about one-fourth of the patients [20]. Thus, more recently, drawing also from the evidence for an autoimmune component in AZOOR, attempts to actively treat AZOOR patients to prevent the development of permanent visual deficits have been made, and success has been reported with steroids administered either

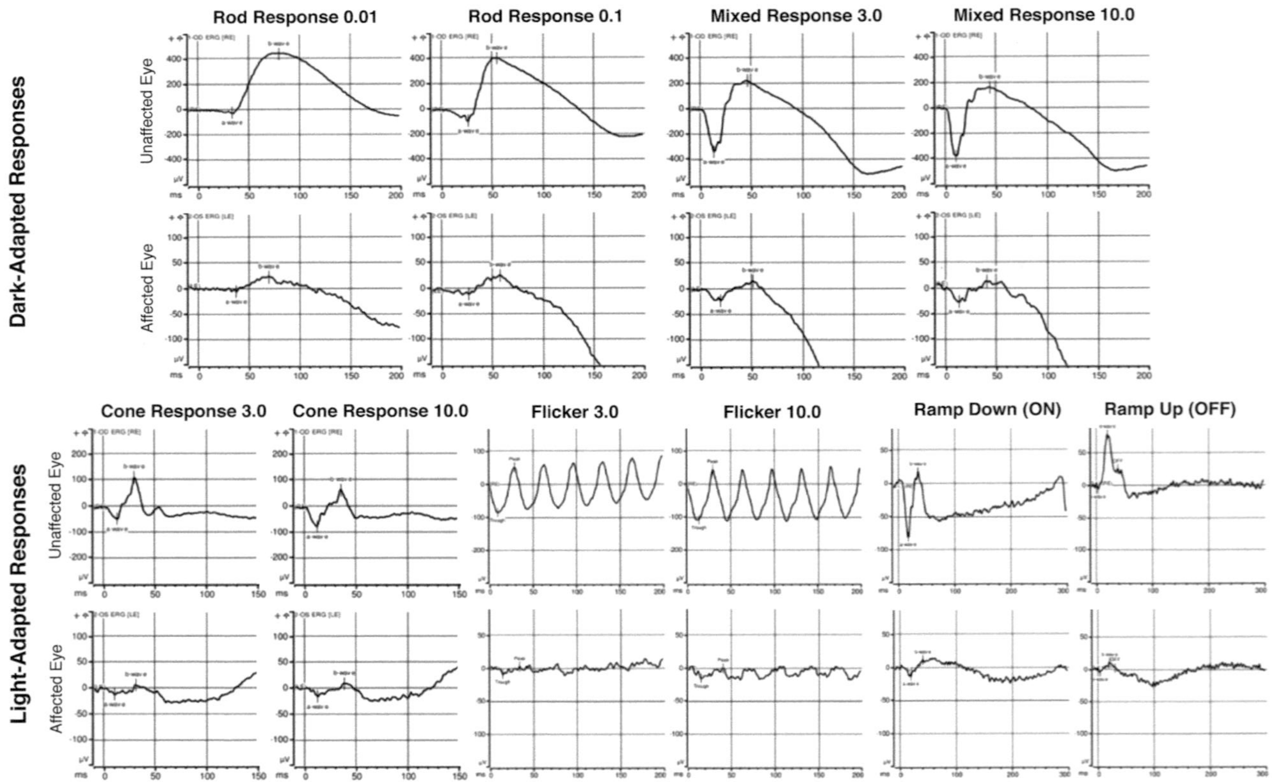

Fig. 8.2 Electroretinographic findings in acute zonal occult outer retinopathy (AZOOR). The ffERG from PT1 reveals a stark difference in all responses between the right (unaffected) and the left eye (affected). All responses are markedly smaller in the left eye (note the difference in the amplitude scale for this eye), and there is a mild electronegativity of the mixed responses in this eye compared with the fellow eye. The flicker response is especially reduced and delayed. Both the ON- and OFF-bipolar cells response to ramp stimuli are markedly reduced, delayed, and disrupted morphologically as well

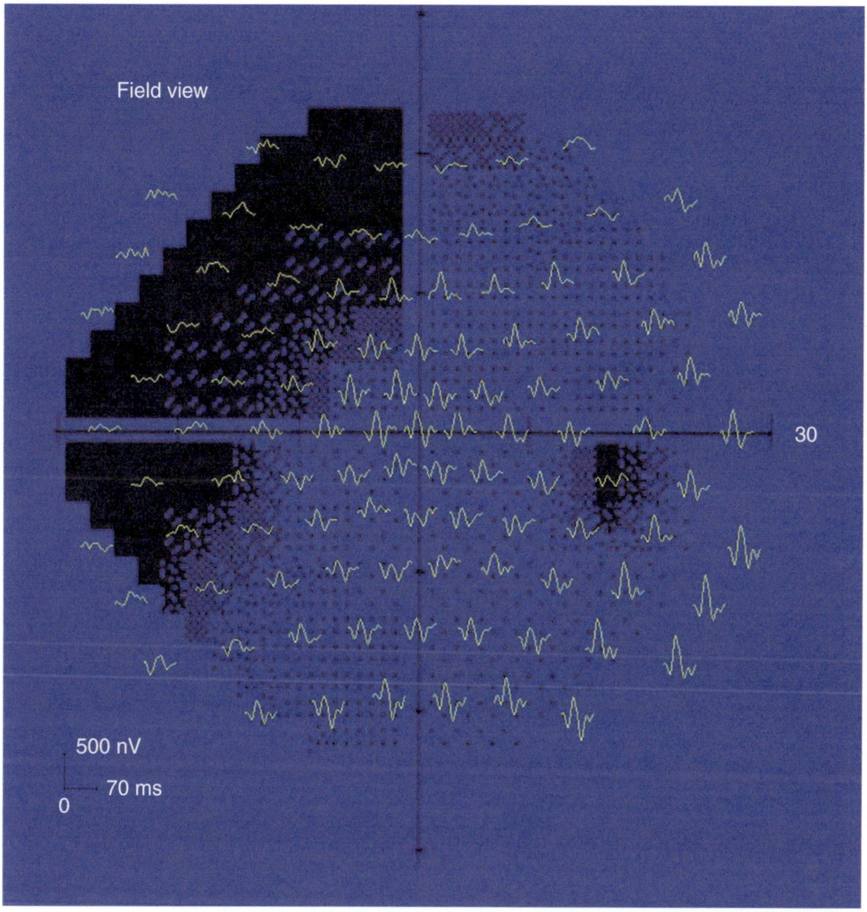

Fig. 8.3 Perimetric and multifocal ERG findings in acute zonal occult outer retinopathy (AZOOR). Multifocal ERGs (yellow traces) superimposed on Humphrey 30–2 central visual field plot of the right eye of a 17-year-old male patient with acute monocular vision loss associated with a viral prodrome and was diagnosed with AZOOR. The overlay shows mfERG abnormalities matching the visual field loss experienced by the patient

systemically or intravitreally [32–34]. In one case of presumed AZOOR that had not been responsive to steroids, marked visual recovery was attained with the anti-TNF-α adalimumab (Humira) [34]. Thus, permanent visual sequelae in AZOOR may be potentially avoided by formulating an early correct diagnosis and instituting an immunosuppressive treatment. These initial results, however, very encouraging and evidence-driven, may require further confirmation.

Multiple Evanescent White Dot Syndrome

Multiple evanescent white dot syndrome (MEWDS) was first described in 1984 [35], involves acute, painless unilateral reduction of visual acuity as low as 20/200, visual field loss, fluorescein leakage from disc capillaries and staining in retinal pigment epithelium (RPE), peripapillary hypofluorescence in indocyanine green angiography, photopsias, floaters, and dyschromatopsia [36–40]. The ocular fundus shows multiple white dots with diameter of 100–200 μm at RPE or the outer retina, and RPE granularity that usually spare the fovea, and are often subclinical and best appreciated on FAF imaging (Fig. 8.4a, b). Optical coherence tomography (OCT) shows disruption of the outer retinal layers in MEWDS (Fig. 8.4c). Most of the MEWDS patients are female, from about 15 to 50 years old. About one-third to half of MEWDS patients suffer from flu-like prodromes. This disease is self-limiting, improves spontaneously, and no treatment is usually needed. Most MEWDS patients regain good visual acuity in 3–9 weeks. Photopsias and scotomata gradually disappear. The fundus lesions are gradually replaced by mild pigment mottling or chorioretinal scarring (Fig. 8.4). Recurrence occurs in about 10% of MEWDS patients. Some MEWDS patients may have persistent blind spot enlargement [36–40].

The differential diagnosis of MEWDS includes acute idiopathic blind spot enlargement syndrome (AIBSES), acute macular neuroretinopathy (AMN), AZOOR, multifocal choroiditis and panuveitis (MCP), punctate inner choroidopathy (PIC), and sarcoidosis. Some of these diseases, such as AIBSES and AMN, can be associated with MEWDS [36–40].

Some of the following electrophysiologic tests can be selected to assist the differential diagnosis according to the results of other examinations. In MEWDS, mfERG N1 amplitudes increase in day 1 to day 7 of the onset of the disease and decrease after 2 weeks of the onset of the MEWDS. The implicit times of mfERG are normal in the course of the disease [41, 42]. OPs are abnormal in some MEWDS patients, reduced in the affected eye with normal scotopic, maximum, photopic, and flicker ERG responses. The multifocal ERG OP amplitudes reduce in the retinal area of the affected eye with normal mfERG N1 and P1 amplitudes [43]. The ratio of the S-cone ERG b-wave amplitude of the affected eyes to that of the normal fellow eyes is significantly lower than the corresponding ratio for the L- and M-cone ERG [44]. The EOG of MEWDS is abnormal. The electrophysiologic tests can recover to normal after the symptoms of MEWDS abate in several weeks [42].

Acute Posterior Multifocal Placoid Pigment Epitheliopathy (APMPPE)

Acute posterior multifocal placoid pigment epitheliopathy (APMPPE) is an inflammatory chorioretinopathy, first described by Gass in 1968 [45], and characterized by multiple yellow-white placoid subretinal lesions of the posterior pole (Fig. 8.5).

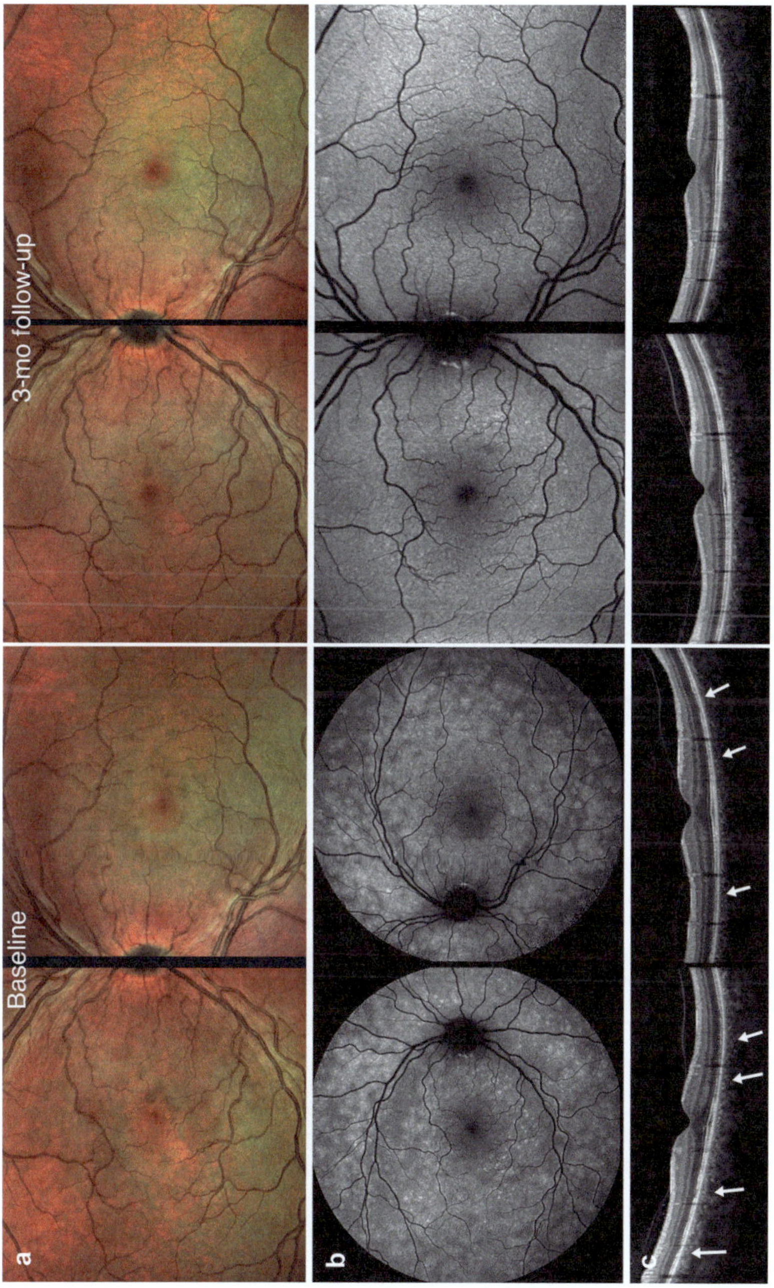

Fig. 8.4 Clinical and imaging examples of MWEDS. This 37-year-old female patient has an acute pericentral visual disturbance. (**a**) Fundus appearance at baseline and 3 months later. Scattered faint deep pigmentary changes and coarse RPE mottling can be seen at baseline, which resolved spontaneously 3 months later. (**b**) The fundus autofluorescence images are much more informative, revealing disseminated spots of hyper-autofluorescence throughout the posterior pole but sparing the fovea. At 3 months, most changes are gone, replaced by fine, punctate, fovea-sparing hyper-autofluorescent sequelae. (**c**) SD-OCT scans show, at baseline, focal disturbances at the RPE/photoreceptor junction (arrows) that largely resolved at 3 months (Images courtesy of Dr. Dilraj Grewal, MD, Duke Eye Center)

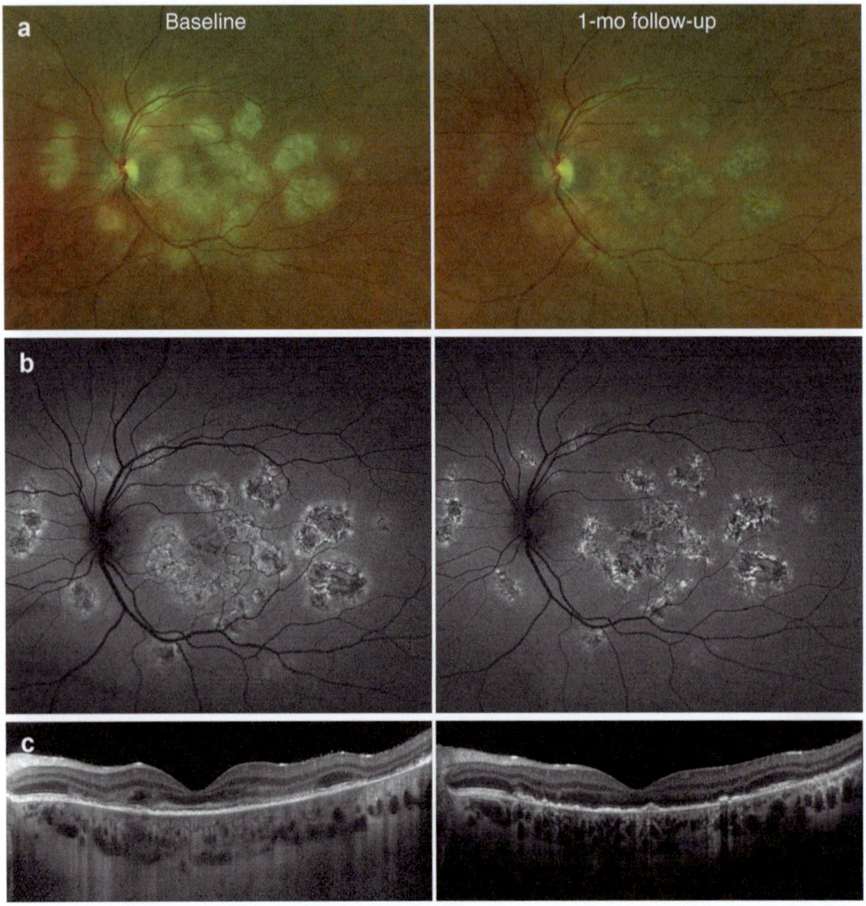

Fig. 8.5 Clinical and imaging example of APMPPE. This 29-year-old female patient with an acute bilateral pericentral visual disturbance, encroaching the center of her vision including the left eye (20/125 in this eye). (**a**) Left fundus appearance at baseline and 1 month later. Disseminated patches of deep yellowish changes can be seen initially, involving also the foveal area, regressing in part already within 1 month. (**b**) Fundus autofluoscence imaging shows disseminated spots of coalescent patches of speckled hypo- and hyper-autofluorescence, partially resolved 1 month later with disseminated sequelae of pigmentary mottling. The visual acuity has spontaneously recovered to 20/25 in this eye. (**c**) SD-OCT scans show, at baseline, focal disseminated areas of thinning of the RPE/photoreceptor junction and development of a thick central band of hyporeflective signal at the interdigitation zone between the RPE and the EZ. There are also scattered hyper-reflectivities overlying the areas of RPE and EZ damage. All these findings appear much improved at 1 month (Images courtesy of Dr. Dilraj Grewal, MD, Duke Eye Center)

APMPPE affects women and men between 20 and 50 years old. APMPPE is often associated with systemic conditions. The patients usually have a rapid onset of blurred vision with central or paracentral scotomas. Visual acuity can decrease to the range of 20/40 to count-fingers. Photopsias can occur before the vision loss. The symptoms usually occur in both eyes but may manifest in each eye in several days apart. Multiple bilateral yellow-white placoid lesions of 1–2 disc diameters at the level of RPE and choroid can be found throughout the fundus from posterior to the equator. The lesions are gradually replaced by RPE atrophy or hyperpigmentation (Fig. 8.5). New lesions may appear in the periphery up to 3 weeks after the onset of APMPPE [46–48].

APMPPE is usually self-delimiting (Fig. 8.5) and resolve in 4–8 weeks (up to 6 months in some patients). The majority of APMPPE patients can recover their visual acuity to 20/40 or better, while foveal involvement is related to a worse prognosis [49]. Approximately one-third of APMPPE patients have a preceding viral or flu-like illness before the onset of APMPPE [50]. This disease may be associated with retinal vasculitis, optic disc edema, subhyaloid hemorrhage, choroidal neovascular membrane formation, and vein occlusion [48, 51].

The etiology of APMPPE is unknown. However, some believe that it is secondary to a delayed-type hypersensitivity vasculitis, due to the inflammation at the level of the choriocapillaris resulting in hypoperfusion and ischemia of the RPE and photoreceptors, causing RPE atrophy and hyperpigmentation in later stage [52, 53]. Genetics may be related to the risk for APMPPE associated with HLA-B7 and HLA-DR2 genetic haplotypes. Viral infection may be another cause of APMPPE [54, 55].

From the electrophysiological standpoint, it has been reported that in APMPPE, the ffERG a- and b-wave amplitudes are slightly decreased in the acute phase [56]. However, the mfERG can show significant decrease in response density in the acute phase and recovery of the amplitudes after the disease reaches the scar stage [57]. Deutman et al. studied EOG and ERG recordings in APMPPE patients. EOG responses were recorded as highly abnormal in the acute phase of the disease, consistent with a severe and widespread dysfunction of the RPE, whereas in the scar stage, the EOG recordings were significantly improved compared to baseline. Thus, the EOG is perhaps the most specific and diagnostic test of all to establish the diagnosis of APMPPE and assess the recovery process after the acute phase.

Birdshot Chorioretinopathy

Birdshot chorioretinopathy was first described in 1949 by Drs. Franceschetti and Bable, when it was first described as "Candle Wax Spot Chorioretinopathy" [58–60]. Subsequently, the terms birdshot chorioretinopathy and vitiliginous chorioretinitis were first used by Drs. Ryan and Maumenee in 1980 and by Dr. Gass in 1981, respectively [58, 60]. While the currently most used term is birdshot chorioretinopathy (or retinochoroidopathy), all three descriptive terms illustrate well the main ophthalmoscopic characteristic of the disease, an acute-onset, relapsing, chronic, bilateral, posterior uveitis with characteristic yellow-white lesions of the fundus. Affected patients most often complain of decreased visual acuity (68%), floaters (29%), nyctalopia (25%), dyschromatopsia (20%), glare (19%), and photopsia (17%) [61].

Birdshot chorioretinopathy is relatively common, accounting for 1–2% of all types of uveitis. It affects primarily Caucasian females between the age of 40 and 60 years old and is strongly associated with a genetic risk factor, HLA-A29, where 80–98% of patients are positive (vs. 7% in the general population) [62]. Despite this clear association, the etiology of this disease remains in part unclear, as it has been hypothesized that the disease may be due to an autoimmune response against arrestin [25].

The term "birdshot" comes from the pattern of the lesions, resembling the gunshot spatter of birdshot. The chorioretinal lesions of birdshot chorioretinopathy are usually approximately 0.25–0.50 disc diameters in size, most often clustered around the optic nerve and the posterior pole, radiating toward the periphery, and most commonly involve the inferonasal peripapillary area (Fig. 8.6). The distribution of the lesions across the fundus can be diffuse, can involve predominantly the macula (or, in fact, spare the macula altogether), and are often asymmetric between the two eyes in both location and amount [25, 58, 60, 61].

In its early (acute) stage, the disease is usually characterized by retinal vascular leakage at or away from the optic disc and by the presence of hypofluorescent lesions on ICG angiography that is usually the optimal imaging test to use at this stage of the disease. In subsequent phases, the so-called middle stage, the condition becomes characterized by prominent birdshot lesions, those are noticeable ophthalmoscopically, by FAF, and fluorescein angiography. In the late stages of the disease, cystoid macular edema (seen in 84% of birdshot chorioretinopathy) [25, 58, 60, 61], disc leakage, RPE changes, retinal atrophy, vascular attenuation, and optic nerve atrophy and subretinal neovascularization can develop and contribute significantly to the vision loss of the birdshot chorioretinopathy patient. Diagnostic criteria have been set at an international workshop [63], and these criteria are summarized in Table 8.1.

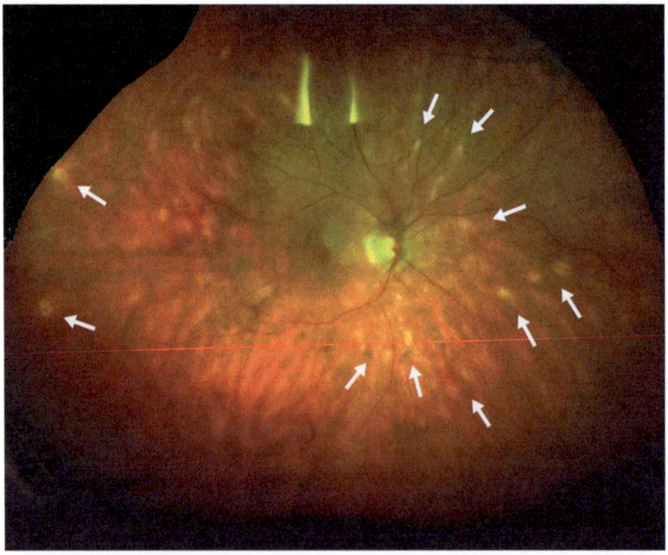

Fig. 8.6 Clinical example of birdshot chorioretinopathy. Typical disseminated faint "candle wax-" like depigmented spots near and around the disc and also in the far temporal periphery characterize this patient with birdshot chorioretinopathy. Subtler findings were seen in the left eye

Table 8.1 Diagnostic criteria for birdshot chorioretinopathy

Required characteristics
Disease in both eyes
≥ 3 peripapillary birdshot lesions (cream-colored, irregular, or elongated choroidal lesions with long axis radiating from optic disc)
≤ 1+ anterior vitreous cells
≤ 2+ vitreous haze
Supportive findings:
HLA-A29 positive
Retinal vasculitis
Cystoid macular edema
Exclusion criteria:
Keratic precipitates
Posterior synechiae
Other causes (i.e., infectious, neoplastic, paraneoplastic, inflammatory)

Visual electrophysiology testing plays an essential role in the diagnosis and management of birdshot chorioretinopathy, both in recognizing – even when the characteristic lesions have not developed yet – the presence of ffERG abnormalities, the most typical charactertistic is an electronegative mixed response with low b-wave amplitude and normal a-wave amplitude (Fig. 8.7) [60]. The light peak of the EOG and the Arden ratio are also reduced fairly consistently in birdshot chorioretinopathy, perhaps reflecting an inflammatory involvement of the RPE at the chorioretinal interface [60]. Visual field testing also plays an important role in the diagnosis and management of birdshot chorioretinopathy, usually revealing multifocal scotomas, arcuate scotomas, enlarged blind spot, and/or central or centrocecal defects.

Unlike the previously described entities, in which treatment is not usually necessary and spontaneous improvement is the rule, birdshot chorioretinopathy is a type of noninfectious posterior uveitis leading to progressive vision loss if left untreated. Thus, treatment is necessary, and it typically includes oral or intravitreal/subtenon steroids, but the vast majority of affected patients is treated systemically with some other form of immunomodulatory therapy (IMT), with which it has been shown that both visual acuity and visual fields loss can stabilize or improve in as many as 89% of the cases [61, 62, 64]. Success has been reported as of late also in recalcitrant cases with many of the newer biologic IMT agents more recently available [65–74]. Thus, promptly recognizing and correctly diagnosing birdshot chorioretinopathy have important prognostic and therapeutic implications.

Autoimmune Retinopathy and Neuroretinopathy

Autoimmune retinopathy (AIR) and neuroretinopathy (AINR, also known as autoimmune-related retinopathy and optic neuropathy or ARRON) are an acquired group of diseases of the retinal and/or the optic nerve caused by autoantibody (AAb)-

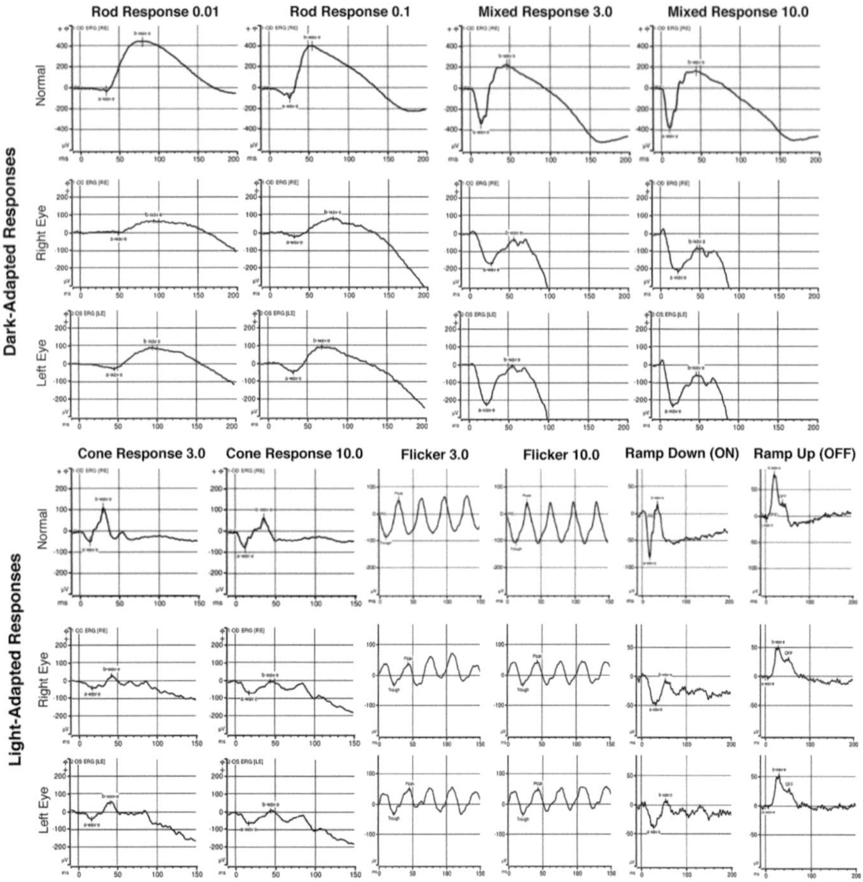

Fig. 8.7 Electroretinographic findings in birdshot chorioretinopathy. The ffERG response of this patient (same as shown in Fig. 8.6) shows bilateral reduction of all responses, which is more severe in rod responses than in cone responses, with marked electronegativity of the mixed responses. The cone-driven responses are also delayed, including the ON and the OFF responses that confirm the ON-bipolar cell compromise in both eyes, but all findings are more severe in the right eye. This pattern of ffERG abnormalities alongside the typical fundus changes is diagnostic for birdshot chorioretinopathy

mediated inflammation and damage to the targeted tissues [17, 18, 29, 75–89]. AIR and AINR/ARRON include two main groups: the paraneoplastic (pAIR/AINR) and non-paraneoplastic (npAIR/AINR) ones. The npAIR forms are considerably more common than the pAIR ones [17, 90, 91]: in our own personal experience, only approximately 20–25% of the cases can be demonstrated to be paraneoplastic. The pAIRs/AINRs can be mainly classified into cancer-associated retinopathy (CAR) and melanoma-associated retinopathy (MAR). A recently recognized pAIR entity is unusual to the extent that it is a special form of CAR, related to a nonmetastatic form of cancer, the basal cell carcinoma-associated retinopathy, and optic neuropathy (BARN) complex. In its fullest expression, this entity also encompasses a significant retinal pigment epitheliopathy, in which case the broader acronym of BARRN has been proposed [31]. Another special case of CAR is the very rare teratoma-associated one [92, 93].

The symptoms of AIR do not differ between npAIR and pAIR except for the latter ones having a plausible temporal connection with a diagnosis of cancer, when this information is available. However, it must be noted that, not infrequently, the visual symptoms can precede the diagnosis (but not the onset) of cancer and, thus, the ophthalmologist and the visual electrophysiologist in particular play a pivotal role in raising the suspicion of a diagnosis of AIR/CAR and spearheading the general health workups in patients who are not yet aware of having a hidden cancer in their body. This can lead to a life-saving discovery for the affected patients. Thus, awareness of these conditions and raising attention to the possible diagnosis on the differential can play a huge role in the well-being and often survival of an affected patient, if the condition turned out to be pAIR.

The manifestations of AIR vary significantly and may include acute to subacute decrease of visual acuity, visual field deficits, photophobia, night blindness, photopsias, and other vague visual disturbances such as seeing wavy "things" in the visual field or other phenomena such as a snowy TV screen and dyschromatopsia [17, 85, 90, 91]. It has been our consistent experience that while the vast majority of AIR patients is affected binocularly, they will most often display asymmetric findings between the two eyes and greater severity in one eye over the other. AIR can be fully monocular too, although usually, to an in-depth evaluation of the patient, some signs of dysfunction will also come to surface in the seemingly unaffected fellow eye even in presumed unilateral cases. With the exception of manifesting female carriers of X-linked RP [94] or other such dystrophies, the vast majority, if not all of the so-called "unilateral RP" cases, reported in the literature should be presumed inflammatory and, most likely, a form of AIR until proven otherwise.

The age of the onset of AIR varies considerably. For the most part, it is an adult-onset problem, and it has been reported that patients with pAIR (CAR) tend to be significantly older than npAIR ones [17, 91]. Childhood-onset cases have been reported [95]. We have confirmed repeatedly both findings in our personal experience. Once the symptoms of AIR begin, especially in the case of npAIR, the progression can be relatively rapid and, again, asymmetric between the two eyes, but it can also be more indolent and chronic, mimicking RP or one of the other retinal dystrophies [17, 85, 90, 91]. The latter cases are particularly challenging to differentiate from true RP, and thus in these cases, it is important to pursue diagnostic tests to address both options.

The clinical manifestations of AIR vary significantly as well. In many patients, especially in early stage, the fundus examination can be completely normal or have only subtle abnormalities so that they can escape a cursory examination of the retina. These are the most difficult cases to diagnose, and it has been our experience that their diagnosis can unfortunately be delayed by years because of the lack of fundus findings. Fortunately, the increasing awareness of these conditions and the nowadays widespread availability of imaging studies such as OCT and, in a more limited fashion, FAF, are rendering the earlier recognition of the organic nature of the complaints patients have, and referrals for suspected AIR or CAR are nowadays much more timely than in the relatively recent past. When observed, fundus findings in AIR patients can include the entire gamut of the manifestations of a retinal degenerative and inflammatory disease (Fig. 8.8), such as bone spicule-like, punctate, or nummular pigment deposits, and RPE abnormalities – again variable from subtle to severe. Both frank retinal and RPE atrophy are possible, and these changes are usually patchy and often distributed in a different pattern distribution between the two eyes. The retinal vasculature can be attenuated in AIR, but it more often exhibits vasculitic changes in association to that, such as sheathings and/or strictly paravascular pigment deposits tracking the larger blood vessels. Macular edema is often seen in AIR and CAR patients, which are the changes at the optic nerve head level, and, as noted above, the other signs of optic neuropathy (see below for further discussion of this particular aspect).

The diagnosis of AIR can be challenging, because demonstrating the presence of the circulating antiretinal AAbs alone, even if performed by an accredited diagnostic laboratory, is not sufficient per se to formulate the diagnosis of AIR or CAR [29, 77, 96, 97]. For example, AAbs directed against macular antigens have been detected in AMD. However, some level of autoreactivity is also observed in unaffected control subjects [98–100]. Thus, any detected reactivity must be interpreted into appropriate context, and the clinical and functional workup of suspected AIR or CAR patients is of paramount importance in the differential diagnostic process and in the interpretation of the outcome of serological testing.

Fig. 8.8 Clinical ophthalmoscopic findings in autoimmune retinopathy. (**a–f**) The commonest ophthalmoscopic chorioretinal manifestations seen in AIR/AINR patients are illustrated in these panels. Typical nummular changes, characterized either by full-thickness punched out chorioretinal lesions (thick arrows) that can be completely depigmented (asterisk in **a**), deeply pigmented, resembling PRP laser scars (asterisks in **b** and **c**), or partially depigmented with a surrounding pigmented ring (thick arrows in **d**). The cases illustrated in **e** and **f** exhibit focal patches of bone spicule-like changes (thin arrows) but some of the pigmentary deposits tracking large vessels suggesting the sequelae of vasculitic events in these cases. In the case in **f**, there are some nummular depigmented changes associated with these focal patchy peripheral changes (thick arrows). (**g–k**) The commonest ophthalmoscopic papillary and peripapillary manifestations seen in AIR/AINR patients are illustrated in these panels. Various degrees of peripapillary atrophy are shown in **g–i**, in **g** also associated with significant pigmentary changes, in **h** mainly associated with depigmentation (small arrows), a parapapillary pigment deposit (asterisk), and focal vascular pigmentary sheathings (small arrowheads). The case in Panel I with an admixture of focal pigmentary (thin arrow) and atrophy (asterisk) alongside fuzzy disc margins and hyperemia nasally (thick arrow). The case in **j** shows pigmentary deposits at the disc margin reminiscent of multifocal juxtapapillary chorioretinitis surrounding all sides of the disc, whereas the one in **k** is notable for frank disc hyperemia. All cases experienced subacute vision loss at various ages, ranging from the late teens to the late 60s, some of the cases were paraneoplastic in nature, and each had evidence of both retinopathy and optic neuropathy (AINR) and positive AAbs consistent with both findings

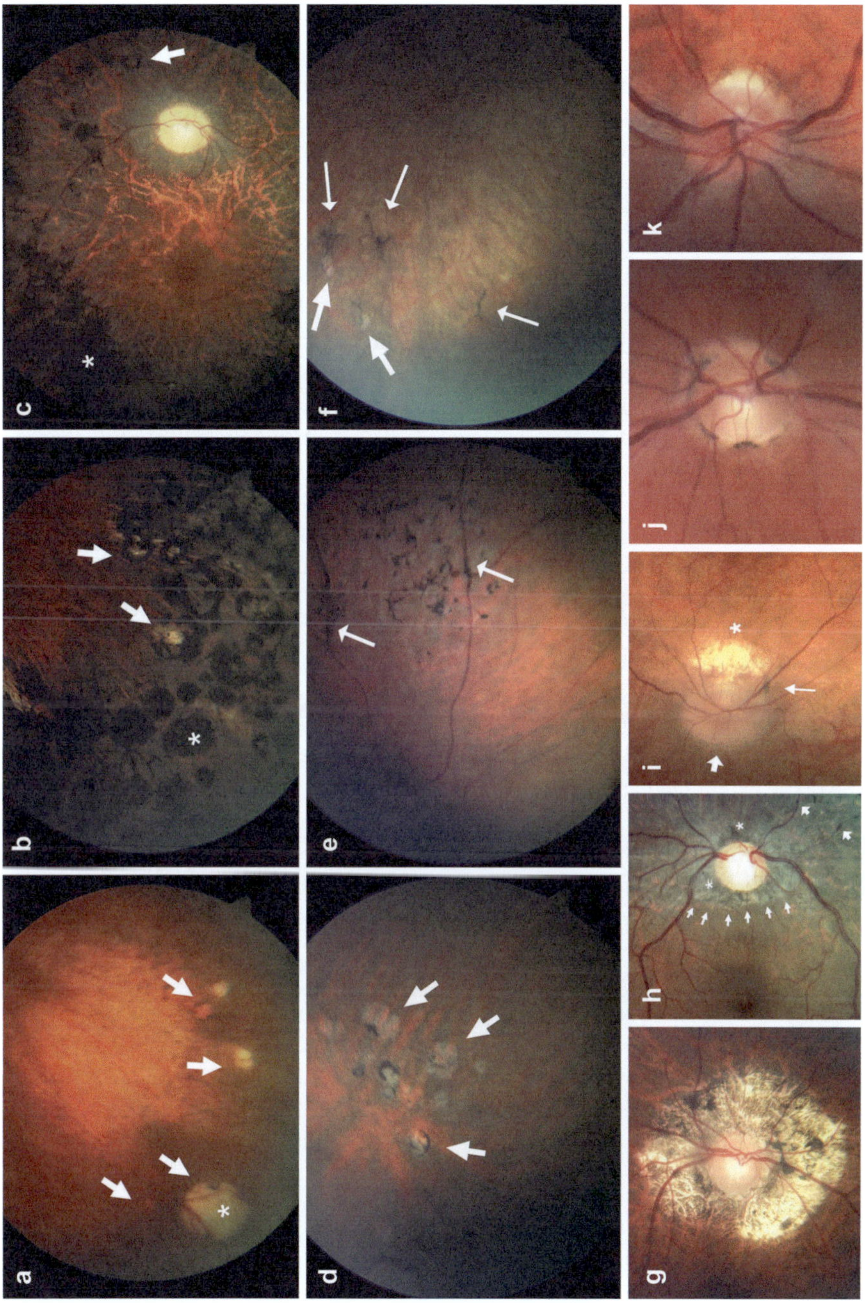
a
b
c
d
e
f
g
h
i
j
k

From a strict retinal standpoint, the ffERG in both pAIR and npAIR usually shows abnormalities of either the rod responses, the cone ones, or both (Fig. 8.9). In addition, there is often evidence for post-receptoral involvement as well, exemplified by electronegative mixed ffERGs and/or evidence of ON- and/or OFF-bipolar cell compromise [76, 91, 101]. Electronegative mixed ffERGs have been reported to be especially common in MAR patients, due to the presence of AAbs directed against bipolar cell antigens [102–104]. However, MAR patients can also have other electroretinographic phenotypes [105], and the electronegative mixed ffERG is truly more specific only of AAbs recognizing bipolar cell antigens, and not any disease in particular. As such, other forms of CAR can display this particular waveform [106]. Furthermore, it has been reported that the ffERG of patients with anti-enolase AAbs tends to show cone-predominant, if not exclusive, functional compromise [76]. Consistent with this finding, most of these patients usually complain of photophobia. The mfERGs in AIR can be informative as well, as it can reveal the decrease of macular cone responses in the central field, providing more evidence for the diagnosis in cases whose ffERG may be only subtly compromised [76, 91, 101]. Lastly, in patients exhibiting anti-bestrophin AAbs, other vitelliform paraneoplastic manifestations, or other forms of autoimmune manifestations linked to widespread RPE compromises, the Arden ratio of the EOG is usually abnormally low. All ffERG, mfERG, and EOG findings usually mirror any existing asymmetry in retinal and/or RPE disease, if the asymmetry in the vision loss is in fact attributable to a retinal etiology.

However, one fundamental problem in the workup of AIR and CAR patients is that approximately 70% of them have serological evidence of anti-optic nerve AAbs, and these findings are matched by evidence of an optic neuropathy from a functional and imaging point of view and at times also by clinical observation [84]. The extent the patients exhibit an "optic neuropathy" is even higher when one takes into account the fact that AAbs recognizing antigens expressed in the retinal ganglion cell (RGCL) or nerve fiber layer (RNFL) will be "read" technically as anti-retinal AAbs both from a diagnostic serology and a retinal immunohistochemistry (IHC) standpoint. However, from a functional standpoint, they will cause an optic neuropathy and either a measurable dysfunction of the optic nerve/visual pathways, microanatomical imaging changes seen on OCT that can and will reveal the neuropathy overlay in these patients (AINR/ARRON if non-paraneoplastic or CARON if cancer-associated), or both. The presence of isolated cancer-associated optic neuropathy has also been well documented [28, 87].

Thus, the workup of the AIR/CAR patient usually needs to also involve direct and indirect measures of the RGCL, RNFL, and optic nerve/pathways health and function (Fig. 8.10) to ensure the full breadth of the manifestations affecting AIR/CAR patients are captured and that potential inconsistencies in the findings (type of symptoms, asymmetry, type of visual field defects) are resolved effectively, allowing the eye care provider and the visual electrophysiologist to raise the possible

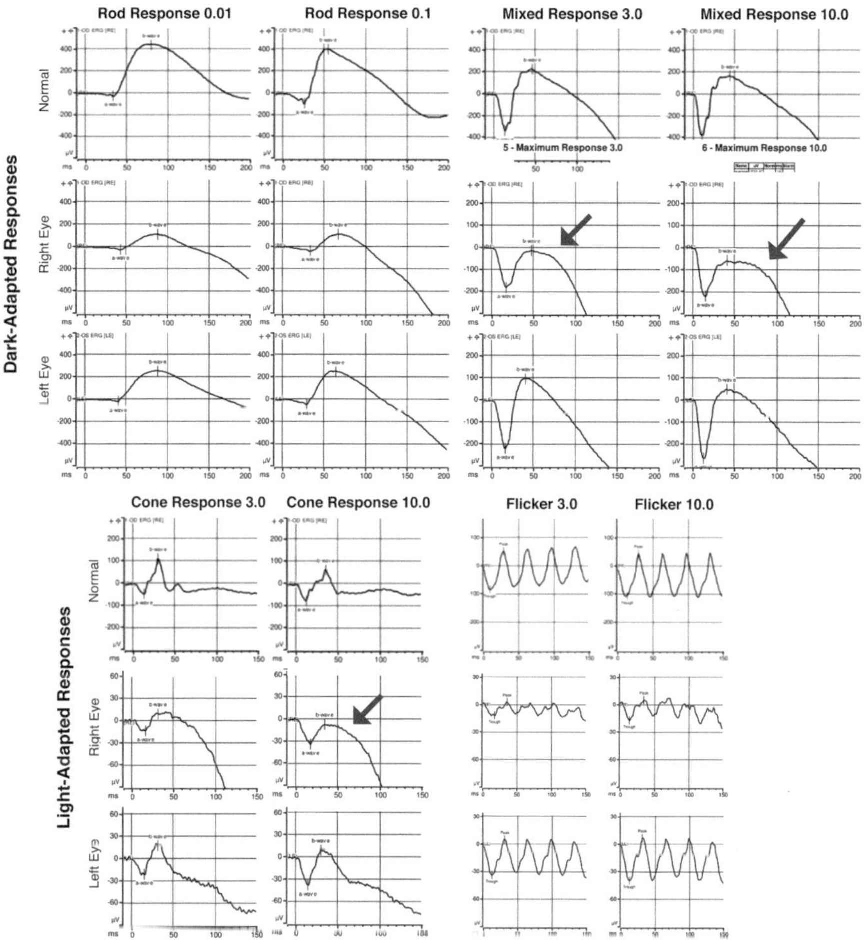

Fig. 8.9 Electroretinographic findings in autoimmune retinopathy. The ffERG of this 80-year-old Caucasian male patient is remarkable for an asymmetric retinopathy, affecting cone responses more than rod responses (note the different scale of the cone responses of the patient). Both of rod and cone responses in the right eye show a marked post-receptoral component (electronegative mixed and cone right ffERGs, dark gray arrows). This patient had a 2-year-old history of subacute profound vision loss, following a 5-year-old history of bladder cancer. Despite the all in all very robust ffERG response preservation and cone-only compromise in the left eye, this patient has count finger vision in each eye, due to a concomitant severe paraneoplastic optic neuropathy. His pattern VEPs were markedly delayed in the left "good eye" (from a ffERG standpoint) and virtually nonrecordable in the right. He tested positive not only for anti-retinal AAbs but also for numerous anti-optic nerve AAbs, accounting well for the severity of his optic nerve compromise, which in his case was mainly retrobulbar. An MRI was also performed, ruling out other intracranial processes that may account for the profound central vision loss

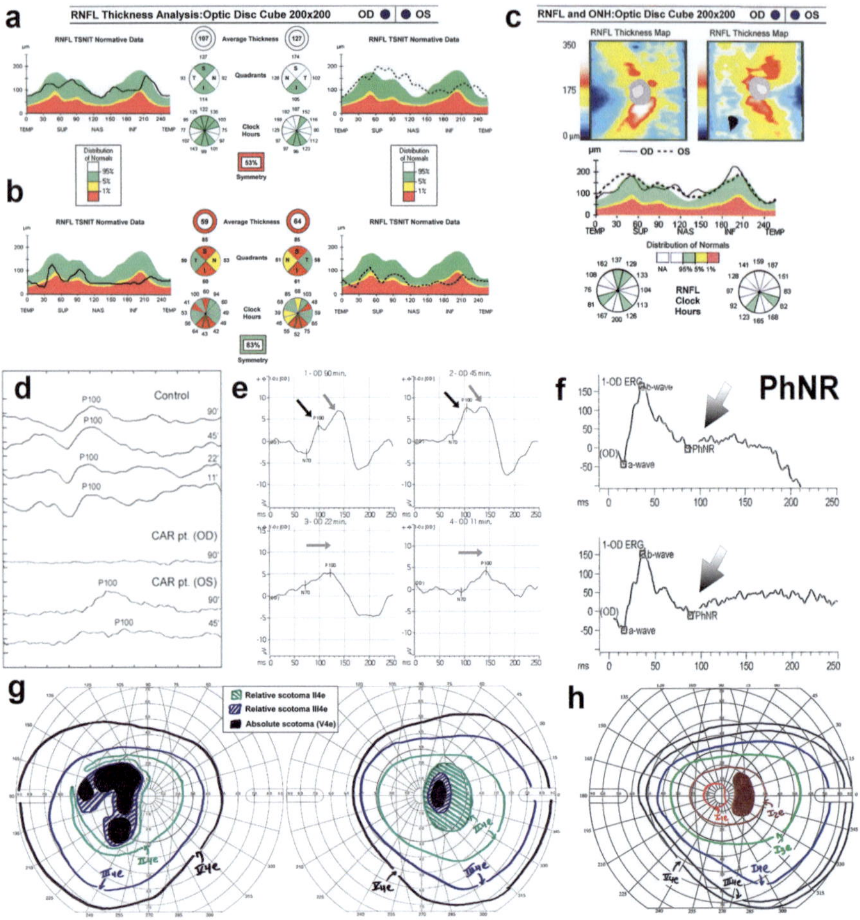

Fig. 8.10 Evidence for optic neuropathy in autoimmune retinopathy. (**a–c**) Peripapillary RNFL analyses of cases of AINR (**a**, **b**) and CARON linked to a newly diagnosed colon cancer discovered after the diagnosis of CARON (**c**) illustrating the range of RNFL anomalies seen in these conditions: (**a**) asymmetric RNFL swelling (almost exclusively in the left eye); (**b**) bilateral RNFL thinning; and (**c**) bilateral but asymmetric RNFL thickening. (**d**) Pattern VEP evidence for asymmetric optic neuropathy (much more in the right eye) in a case of CARON linked to multiple myeloma, explaining a markedly asymmetric visual field loss in the presence of a symmetric retinopathy by ffERG criteria. (**e**) Morphological PVEP response abnormalities with P100 splitting (black and gray arrows) at larger check sizes, followed by increasing P100 delays (gray arrows) to smaller checks. (**f**) In a patient with marked disc pallor, asymmetric and oddly shaped visual field loss, and cone-only response delays of ffERG testing ("cone dysfunction syndrome" type of ffERG anomaly), the RGC-derived PhNR can be localized but its amplitude markedly reduced (gray arrows). This case also exhibited pure acquired blue-yellow axis dyschromatopsia on Panel D15 testing (not shown). (**g–h**) Various patterns of kinetic visual field loss indicative of optic nerve/disc compromise: the fields shown in **g** belong to the patient whose RNFL OCT is shown in **c** (CARON). The enlarged blind spot case shown in Fig. 8.10h belongs to a case with npAINR in which the optic neuropathy is the main driver of the vision loss. Neither one of these cases had peripapillary atrophy or other chorioretinal changes sufficient to explain the findings

diagnosis of AIR/CAR and an appropriate serological workup can follow. In our experience [31, 84, 85, 107, 108], indications a patient may have a concomitant optic neuropathy include the following: Upon color vision testing (with Panel D15 confrontation method or others that can gauge also axes other than the red-green one similar to what the Ishihara plates do) and dyschromatopsia along the blue-yellow axis found in visual field testing are bundle-like or arcuate defects or enlarged blind spot(s) in the absence of retinal/RPE peripapillary changes that may account for optic neuropathy. Additionally, delayed or otherwise morphologically abnormal pattern VEP waveforms can be found even in patients whose acuity may be 20/20 – in our experience. The most common morphological findings include the presence of positive–negative–positive (PNP) complex resulting usually either from marked P100 delay shifting the entire PVEP response over to the right or from P100 peak splits. Most often changes are attributable to bundles of nerve fibers firing asynchronously from one another due to sectoral RGCL/RNFL defects; thinned or, more commonly, thickened RNFL; or other disc changes such as notching and focal pigment deposits seen ophthalmoscopically, reminiscent of a juxtapapillary chorioretinitis. Also observed are changes in the RGCL or RNFL overall thickness across the macular region upon OCT volume scan segmentation analyses. We first provided evidence of this when automated segmentation options did not exist yet [107], and this phenomenon can now often be shown by simply using the built-in algorithms of the imaging software.

Upon ffERG testing, indirect evidence for optic nerve/pathway compromise can come from ascertaining symmetric ffERG losses in patients with significantly asymmetric visual field losses that cannot be accounted for simply by retinal etiology. Additional evidence of RGCL compromise can often be found in these patients by recording the photopic negative response (PhNR), a component of the ffERG originating from RGC-driven responses and that has now become an established measure of RGC health in a number of optic neuropathies and part of the extended ISCEV standard protocol [109–114]. The PhNR is especially helpful in those patients whose central acuity may be too low and/or central scotomas too dense to permit reliable PVEP measurements. Utilization of the PhNR, however, becomes challenging when the extent of the retinal disease in AIR is such that the responses, especially the photopic ones, are profoundly reduced or non-recordable. Patients who exhibit one or more of these features supporting the suspicion of an optic nerve/pathway involvement, either as the sole reason for the autoimmune-mediated vision loss or as an added layer of complexity to a case of AIR (AINR or ARRON) or CAR (CARON), will typically exhibit positive serologies specifically for anti-optic nerve AAbs and/or exhibit positive anti-retinal AAbs with staining pattern on diagnostic retinal IHC affecting the RGCL, the RNFL, or both [31, 84, 87, 108].

Aside from the potential life-saving implications of detecting paraneoplastic vision loss in patients who are not yet aware of having a hidden cancer in their body as the trigger to their visual symptoms, recognition and confirmation of cases of AIR/AINR/ARRON or CAR/CARON has also important implications from a treatment standpoint, because many of these patients, especially if identified sufficiently early in the course of the disease, can respond well to treatment, and further vision loss can be averted, if not partial recovery can be achieved [91]. It has been our experience that the initial benefits of subtenon or intravitreal steroids can be extended beyond the initial "run in phase" and can be used long-term to manage successfully these patients. The local treatment regimen is the treatment of choice in paraneoplastic vision loss, because it has been reported systemic IMT can pose a risk to the cancer patient and potentially facilitate the growth, metastatic potential, and/or recurrence of malignancies. In non-paraneoplastic cases, however, systemic IMT is definitely effective and often necessary. Many instances of documented benefit have been reported with rituximab (Rituxin) [115–120], and recently success has been reported also from other biologic agent-based approaches, such as tocilizumab (Actemra) [121] and adalimumab (Humira) [122], the latter being already indicated for noninfectious posterior inflammatory diseases [123].

References

1. Gass JD. Acute zonal occult outer retinopathy. Donders Lecture: The Netherlands Ophthalmological Society, Maastricht, Holland, June 19, 1992. J Clin Neuroophthalmol. 1993;13(2):79–97.
2. Boudreault KA, et al. Quantitative autofluorescence intensities in acute zonal occult outer retinopathy vs healthy eyes. JAMA Ophthalmol. 2017;135(12):1330–8.
3. Wang Q, et al. Fundus autofluorescence imaging in the assessment of acute zonal occult outer retinopathy. Ophthalmologica. 2017;237(3):153–8.
4. Mrejen S, et al. Acute zonal occult outer retinopathy: a classification based on multimodal imaging. JAMA Ophthalmol. 2014;132(9):1089–98.
5. Fujiwara T, et al. Fundus autofluorescence and optical coherence tomographic findings in acute zonal occult outer retinopathy. Retina. 2010;30(8):1206–16.
6. Takai Y, et al. Morphological study of acute zonal occult outer retinopathy (AZOOR) by multiplanar optical coherence tomography. Acta Ophthalmol. 2009;87(4):408–18.
7. Jacobson SG, et al. Pattern of retinal dysfunction in acute zonal occult outer retinopathy. Ophthalmology. 1995;102(8):1187–98.
8. Francis PJ, et al. Acute zonal occult outer retinopathy: towards a set of diagnostic criteria. Br J Ophthalmol. 2005;89(1):70–3.
9. Li D, Kishi S. Loss of photoreceptor outer segment in acute zonal occult outer retinopathy. Arch Ophthalmol. 2007;125(9):1194–200.
10. Mkrtchyan M, et al. Outer retinal structure in patients with acute zonal occult outer retinopathy. Am J Ophthalmol. 2012;153(4):757–68, 768 e1.
11. Nakao S, et al. Spontaneous remission of acute zonal occult outer retinopathy: follow-up using adaptive optics scanning laser ophthalmoscopy. Graefes Arch Clin Exp Ophthalmol. 2015;253(6):839–43.
12. Makino S, Tampo H. Changes in optical coherence tomography findings in acute zonal occult outer retinopathy. Case Rep Ophthalmol. 2013;4(3):99–104.
13. Tagami M, et al. Autologous antibodies to outer retina in acute zonal occult outer retinopathy. Jpn J Ophthalmol. 2014;58(6):462–72.
14. Jampol LM, Wiredu A. MEWDS, MFC, PIC, AMN, AIBSE, and AZOOR: one disease or many? Retina. 1995;15(5):373–8.
15. Gass JD. The acute zonal outer retinopathies. Am J Ophthalmol. 2000;130(5):655–7.
16. Gass JD. Are acute zonal occult outer retinopathy and the white spot syndromes (AZOOR complex) specific autoimmune diseases? Am J Ophthalmol. 2003;135(3):380–1.

17. Heckenlively JR, Ferreyra HA. Autoimmune retinopathy: a review and summary. Semin Immunopathol. 2008;30(2):127–34.
18. He SX, et al. Auto antibodies against retinal pigment epithelium in patients with atypical retinopathies. Invest Ophthalmol Vis Sci. 2010;51(5):E-Abstract 3771.
19. Qian CX, et al. Prevalence of Antiretinal antibodies in acute zonal occult outer retinopathy: a comprehensive review of 25 cases. Am J Ophthalmol. 2017;176:210–8.
20. Gass JD, Agarwal A, Scott IU. Acute zonal occult outer retinopathy: a long-term follow-up study. Am J Ophthalmol. 2002;134(3):329–39.
21. Abo-Shasha R, et al. Is acute zonal occult outer retinopathy an autoimmune condition? A case report and literature review. Ophthalmic Surg Lasers Imaging Retina. 2015;46(6):662–5.
22. Priem HA, et al. Electrophysiologic studies in birdshot chorioretinopathy. Am J Ophthalmol. 1988;106:430–6.
23. Hirose T, et al. Retinal function in birdshot retinochoroidopathy. Acta Ophthalmol. 1991;69:327–37.
24. Menezo V, Taylor SR. Birdshot uveitis: current and emerging treatment options. Clin Ophthalmol. 2014;8:73–81.
25. Holak HM, Szymaniec S, Holak SA. The pathogenesis of birdshot chorioretinopathy. Surv Ophthalmol. 2006;51(4):446–7; author reply 447.
26. Lei B, et al. Human melanoma-associated retinopathy (MAR) antibodies alter the retinal ON-response of the monkey ERG in vivo. Invest Ophthalmol Vis Sci. 2000;41:262–6.
27. Shiraga S, Adamus G. Mechanism of CAR syndrome: anti-recoverin antibodies are the inducers of retinal cell apoptotic death via the caspase 9- and caspase 3-dependent pathway. J Neuroimmunol. 2002;132(1–2):72–82.
28. Chan JW. Paraneoplastic retinopathies and optic neuropathies. Surv Ophthalmol. 2003;48(1):12–38.
29. Adamus G, Ren G, Weleber RG. Autoantibodies against retinal proteins in paraneoplastic and autoimmune retinopathy. BMC Ophthalmol. 2004;4:5.
30. Lu Y, et al. Melanoma-associated retinopathy: a paraneoplastic autoimmune complication. Arch Ophthalmol. 2009;127(12):1572–80.
31. Iannaccone A, et al. Basal cell carcinoma-associated retinopathy and optic neuropathy (BARN): a novel paraneoplastic entity. Invest Ophthalmol Vis Sci. 2019;60(5):E-Abstract 2780.
32. Barnes AC, et al. Treatment of acute zonal occult outer retinopathy with intravitreal steroids. Ophthalmic Surg Lasers Imaging Retina. 2018;49(7):504–9.
33. Kuo YC, Chen N, Tsai RK. Acute Zonal Occult Outer Retinopathy (AZOOR): a case report of vision improvement after intravitreal injection of Ozurdex. BMC Ophthalmol. 2017;17(1):236.
34. Chen SN, Yang CH, Yang CM. Systemic corticosteroids therapy in the management of acute zonal occult outer retinopathy. J Ophthalmol. 2015;2015:793026.
35. Jampol LM, et al. Multiple evanescent white dot syndrome. I. Clinical findings. Arch Ophthalmol. 1984;102(5):671–4.
36. Raven ML, et al. Multi-modal imaging and anatomic classification of the white dot syndromes. Int J Retina Vitreous. 2017;3:12.
37. dell'Omo R, Pavesio CE. Multiple evanescent white dot syndrome (MEWDS). Int Ophthalmol Clin. 2012;52(4):221–8.
38. Ryan PT. Multiple evanescent white dot syndrome: a review and case report. Clin Exp Optom. 2010;93(5):324–9.
39. Matsumoto Y, Haen SP, Spaide RF. The white dot syndromes. Compr Ophthalmol Updat. 2007;8(4):179–200; discussion 203–4.
40. Quillen DA, et al. The white dot syndromes. Am J Ophthalmol. 2004;137(3):538–50.
41. Feigl B, Haas A, El-Shabrawi Y. Multifocal ERG in multiple evanescent white dot syndrome. Graefes Arch Clin Exp Ophthalmol. 2002;240(8):615–21.

42. Chen D, Martidis A, Baumal CR. Transient multifocal electroretinogram dysfunction in multiple evanescent white dot syndrome. Ophthalmic Surg Lasers. 2002;33(3):246–9.
43. Cheng JY, et al. The outer and inner retinal function in patients with multiple evanescent white dot syndrome. Clin Exp Ophthalmol. 2009;37(5):478–84.
44. Yamamoto S, et al. S-cone electroretinograms in multiple evanescent white dot syndrome. Doc Ophthalmol. 2003;106(2):117–20.
45. Gass JD. Acute posterior multifocal placoid pigment epitheliopathy. Arch Ophthalmol. 1968;80(2):177–85.
46. Roberts TV, Mitchell P. Acute posterior multifocal placoid pigment epitheliopathy: a long-term study. Aust N Z J Ophthalmol. 1997;25(4):277–81.
47. Vianna R, et al. Natural history and visual outcome in patients with APMPPE. Bull Soc Belge Ophtalmol. 1993;248:73–6.
48. de Laey JJ. [Placoid epitheliopathy and serpiginous choroidopathy]. Bull Soc Belge Ophtalmol. 1989;230:105–22.
49. Xerri O, et al. Untreated acute posterior multifocal placoid pigment epitheliopathy (APMPPE): a case series. BMC Ophthalmol. 2018;18(1):76.
50. Brydak-Godowska J, et al. Ocular complications in influenza virus infection. Ocul Immunol Inflamm. 2019;27(4):545–50.
51. Mangeon M, et al. Multimodal evaluation of patients with acute posterior multifocal placoid pigment epitheliopathy and serpiginous choroiditis. Ocul Immunol Inflamm. 2018;26(8):1212–8.
52. Goen TM, Terry JE. Acute posterior multifocal placoid pigment epitheliopathy. J Am Optom Assoc. 1987;58(2):112–7.
53. Cozubas R, et al. Similarities and differences between three different types of white dot syndrome and the therapeutic possibilities. Rom J Ophthalmol. 2018;62(3):183–7.
54. Thomson SP, Roxburgh ST. Acute posterior multifocal placoid pigment epitheliopathy associated with adenovirus infection. Eye (Lond). 2003;17(4):542–4.
55. Azar P Jr, et al. Acute posterior multifocal placoid pigment epitheliopathy associated with an adenovirus type 5 infection. Am J Ophthalmol. 1975;80(6):1003–5.
56. Daniele S, et al. Progression of choroidal atrophy in acute posterior multifocal placoid pigment epitheliopathy. Ophthalmologica. 1998;212(1):66–72.
57. Aoyagi R, et al. Multifocal electroretinographic evaluation of macular function in acute posterior multifocal placoid pigment epitheliopathy. Doc Ophthalmol. 2013;126(3):253–8.
58. Leung E, et al. Birdshot retinochoroidopathy. In: Karth PA, O'Keefe GD, editors. eyewiki.aao.org. American Academy of Ophthalmology; 2019.
59. Franceschetti A, Babel J. La chorioretinite en taches de bougie, manifestation de la maladie de Besnier-Boeck. Ophthalmologica. 1949;118(4–5):701–10.
60. Gasch AT, Smith JA, Whitcup SM. Birdshot retinochoroidopathy. Br J Ophthalmol. 1999;83(2):241–9.
61. Rothova A, et al. Birdshot chorioretinopathy: long-term manifestations and visual prognosis. Ophthalmology. 2004;111(5):954–9.
62. Kiss S, et al. Long-term follow-up of patients with birdshot retinochoroidopathy treated with corticosteroid-sparing systemic immunomodulatory therapy. Ophthalmology. 2005;112(6):1066–71.
63. Levinson RD, et al. Research criteria for the diagnosis of birdshot chorioretinopathy: results of an international consensus conference. Am J Ophthalmol. 2006;141(1):185–7.
64. Thorne JE, et al. Loss of visual field among patients with birdshot chorioretinopathy. Am J Ophthalmol. 2008;145(1):23–8.
65. Leclercq M, et al. Tocilizumab for the treatment of birdshot uveitis that failed interferon alpha and anti-tumor necrosis factor-alpha therapy: two cases report and literature review. Clin Rheumatol. 2018;37(3):849–53.
66. Steeples LR, et al. Adalimumab in refractory cystoid macular edema associated with birdshot chorioretinopathy. Int Ophthalmol. 2018;38(3):1357–62.

67. Lopalco G, et al. IL-6 blockade in the management of non-infectious uveitis. Clin Rheumatol. 2017;36(7):1459–69.
68. Mesquida M, et al. Twenty-four month follow-up of tocilizumab therapy for refractory uveitis-related macular edema. Retina. 2018;38(7):1361–70.
69. Calvo-Rio V, et al. Efficacy of anti-IL6-receptor tocilizumab in refractory cystoid macular edema of birdshot retinochoroidopathy report of two cases and literature review. Ocul Immunol Inflamm. 2017;25(5):604–9.
70. Papo M, et al. Tocilizumab in severe and refractory non-infectious uveitis. Clin Exp Rheumatol. 2014;32(4 Suppl 84):S75–9.
71. Mesquida M, et al. Long-term effects of tocilizumab therapy for refractory uveitis-related macular edema. Ophthalmology. 2014;121(12):2380–6.
72. Adan A, et al. Tocilizumab treatment for refractory uveitis-related cystoid macular edema. Graefes Arch Clin Exp Ophthalmol. 2013;251(11):2627–32.
73. Artornsombudh P, et al. Infliximab treatment of patients with birdshot retinochoroidopathy. Ophthalmology. 2013;120(3):588–92.
74. Sobrin L, et al. Daclizumab for treatment of birdshot chorioretinopathy. Arch Ophthalmol. 2008;126(2):186–91.
75. Magrys A, et al. The role of anti-alpha-enolase autoantibodies in pathogenicity of autoimmune-mediated retinopathy. J Clin Immunol. 2007;27(2):181–92.
76. Weleber RG, et al. Clinical and electrophysiologic characterization of paraneoplastic and autoimmune retinopathies associated with antienolase antibodies. Am J Ophthalmol. 2005;139(5):780–94.
77. Ren G, Adamus G. Cellular targets of anti-alpha-enolase autoantibodies of patients with autoimmune retinopathy. J Autoimmun. 2004;23(2):161–7.
78. Adamus G. Autoantibody-induced apoptosis as a possible mechanism of autoimmune retinopathy. Autoimmun Rev. 2003;2(2):63–8.
79. Adamus G. Antirecoverin antibodies and autoimmune retinopathy. Arch Ophthalmol. 2000;118(11):1577–8.
80. Mizener JB, et al. Autoimmune retinopathy in the absence of cancer. Am J Ophthalmol. 1997;123(5):607–18.
81. Heckenlively JR, et al. Autoimmune retinopathy: patients with antirecoverin immunoreactivity and panretinal degeneration. Arch Ophthalmol. 2000;118(11):1525–33.
82. Adamus G, Karren L. Autoimmunity against carbonic anhydrase II affects retinal cell functions in autoimmune retinopathy. J Autoimmun. 2009;32(2):133–9.
83. Adamus G, Yang S, Weleber RG. Unique epitopes for carbonic anhydrase II autoantibodies related to autoimmune retinopathy and cancer-associated retinopathy. Exp Eye Res. 2016;147:161.
84. Adamus G, et al. Diversity in autoimmunity against retinal, neuronal, and axonal antigens in acquired neuro-retinopathy. J Ophthalmic Inflamm Infect. 2011;1(3):111–21.
85. Radhakrishnan SS, et al. Patterns of visual function loss in autoimmune neuro-retinopathy (AINR): psychophysical and electrophysiological findings [ARVO abstract]. Invest Ophthalmol Vis Sci. 2010;51(5):E-abstract 3547.
86. Iyer SS, et al. Presentation of autoimmune neuro-retinopathy (AINR) patients with anti-carbonic anhydrase II auto-antibodies [ARVO Abstract]. Invest Ophthalmol Vis Sci. 2011;52(6):E-abstract 2924.
87. Carboni G, et al. Bilateral paraneoplastic optic neuropathy and unilateral retinal compromise in association with prostate cancer: a differential diagnostic challenge in a patient with unexplained visual loss. Doc Ophthalmol. 2012;125:63.
88. Epstein RS, et al. Clinical, functional, and imaging characteristics of cancer-associated retinopathy and optic neuropathy. In: American Academy of Ophthalmology meeting. 2014: Chicago, 18–22 Oct 2014.
89. Grange L, et al. Autoimmune retinopathy. Am J Ophthalmol. 2014;157(2):266–272 e1.

90. Comlekoglu DU, Thompson IA, Sen HN. Autoimmune retinopathy. Curr Opin Ophthalmol. 2013;24(6):598–605.
91. Ferreyra HA, et al. Management of autoimmune retinopathies with immunosuppression. Arch Ophthalmol. 2009;127(4):390–7.
92. Turaka K, et al. Carcinoma-associated retinopathy in a young teenager with immature teratoma of the ovary. J AAPOS. 2014;18(4):396–8.
93. Suhler EB, et al. Presumed teratoma-associated paraneoplastic retinopathy. Arch Ophthalmol. 2003;121(1):133–7.
94. Jacobson SG, et al. Interocular asymmetry of visual function in heterozygotes of X-linked retinitis pigmentosa. Exp Eye Res. 1989;48:679–91.
95. Ko AC, et al. Anti-gamma-enolase autoimmune retinopathy manifesting in early childhood. Arch Ophthalmol. 2010;128(12):1590–5.
96. Polans AS, et al. Recoverin, a photoreceptor-specific calcium-binding protein, is expressed by the tumor of a patient with cancer-associated retinopathy. Proc Natl Acad Sci U S A. 1995;92(20):9176–80.
97. Pratesi F, et al. Autoantibodies specific for alpha-enolase in systemic autoimmune disorders. J Rheumatol. 2000;27(1):109–15.
98. Iannaccone A, et al. Autoimmunity in age-related macular degeneration: a possible role player in disease development and progression. Adv Exp Med Biol. 2012;723:11–6.
99. Iannaccone A, et al. Circulating autoantibodies in age-related macular degeneration recognize human macular tissue antigens implicated in autophagy, immunomodulation, and protection from oxidative stress and apoptosis. PLoS One. 2015;10(12):e0145323.
100. Iannaccone A, et al. Retinal pigment epithelium and microglia express the CD5 antigen-like protein, a novel autoantigen in age-related macular degeneration. Exp Eye Res. 2017;155:64–74.
101. Chang PY, Yang CH, Yang CM. Cancer-associated retinopathy in a patient with hepatocellular carcinoma: case report and literature review. Retina. 2005;25(8):1093–6.
102. Andreasson S, Ponjavic V, Ehinger B. Full-field electroretinogram in a patient with cutaneous melanoma-associated retinopathy. Acta Ophthalmol (Copenh). 1993;71:487–90.
103. Milam AH, et al. Autoantibodies against retinal bipolar cells in cutaneous melanoma-associated retinopathy. Invest Ophthalmol Vis Sci. 1993;34:91–100.
104. Lei B, et al. Human melanoma-associated retinopathy (MAR) antibodies alter the retinal ON-response of the monkey ERG in vivo. Invest Ophthalmol Vis Sci. 2000;41(1):262–6.
105. Potter MJ, et al. Autoantibodies to transducin in a patient with melanoma-associated retinopathy. Am J Ophthalmol. 2002;134(1):128–30.
106. Jacobson DM, Adamus G. Retinal anti-bipolar cell antibodies in a patient with paraneoplastic retinopathy and colon carcinoma. Am J Ophthalmol. 2001;131(6):806–8.
107. Birg A, et al. Spectral-domain optical coherence tomography (SD-OCT) findings in patients with autoimmune neuro-retinopathy (AINR). Invest Ophthalmol Vis Sci. 2011;52(5):–E-Abstract 3690.
108. Iannaccone A, et al. Optic nerve/retinal ganglion cell involvement in autoimmune retinopathies resulting in autoimmune neuroretinopathy. In: American Academy of Ophthalmology meeting. Chicago; 2014. p. PO547.
109. Frishman L, et al. ISCEV extended protocol for the photopic negative response (PhNR) of the full-field electroretinogram. Doc Ophthalmol. 2018;136(3):207–11.
110. Majander A, et al. The pattern of retinal ganglion cell dysfunction in Leber hereditary optic neuropathy. Mitochondrion. 2017;36:138–49.
111. Karanjia R, et al. The photopic negative response: an objective measure of retinal ganglion cell function in patients with Leber's hereditary optic neuropathy. Invest Ophthalmol Vis Sci. 2017;58(6):BIO300–6.
112. Katagiri S, et al. Retinal structure and function in eyes with optic nerve hypoplasia. Sci Rep. 2017;7:42480.

113. You Y, et al. Optic neuropathies: characteristic features and mechanisms of retinal ganglion cell loss. Rev Neurosci. 2013;24(3):301–21.
114. Machida S. Clinical applications of the photopic negative response to optic nerve and retinal diseases. J Ophthalmol. 2012;2012:397178.
115. Boudreault K, et al. Efficacy of rituximab in non-paraneoplastic autoimmune retinopathy. Orphanet J Rare Dis. 2017;12(1):129.
116. Forooghian F, MacDonald IM. Rituximab for the treatment of autoimmune retinopathy. Am J Ophthalmol. 2017;180:xv–xvi.
117. Davoudi S, et al. Outcomes in autoimmune retinopathy patients treated with rituximab. Am J Ophthalmol. 2017;180:124–32.
118. Maleki A, et al. Rituximab as a monotherapy or in combination therapy for the treatment of non-paraneoplastic autoimmune retinopathy. Clin Ophthalmol. 2017;11:377–85.
119. Uludag G, et al. Electroretinographic improvement after rituximab therapy in a patient with autoimmune retinopathy. Am J Ophthalmol Case Rep. 2016;2:4–7.
120. Fox A, et al. Rituximab treatment for nonparaneoplastic autoimmune retinopathy. Can J Ophthalmol. 2015;50(6):e101–4.
121. Finn AP, Keenan RT, Jaffe GJ. Reconstitution of the ellipsoid zone with tocilizumab in autoimmune retinopathy. Retin Cases Brief Rep. 2018.
122. Grewal DS, Jaffe GJ, Keenan RT. Sarilumab for recalcitrant cystoid macular edema in non-paraneoplastic autoimmune retinopathy. Retin Cases Brief Rep. 2019.
123. Jaffe GJ, et al. Adalimumab in patients with active noninfectious uveitis. N Engl J Med. 2016;375(10):932–43.

Chapter 9
Characteristics of Visual Electrophysiology in Retinal Toxicities

Minzhong Yu, Alfonso Senatore, Alessandro Iannaccone, Wajiha Jurdi Kheir, and Donnell Creel

Vigabatrin

Vigabatrin (VGB), sold under the brand name Sabril, is an anticonvulsant medication inhibiting gamma-aminobutyric acid (GABA) degradation for the treatment of epilepsy. It is an irreversible inhibitor of GABA transaminase. By irreversibly binding to GABA-transaminase, thus metabolism of GABA is hurdled. The accumulation of GABA then damages the neurons in the eye and brain. It can reduce visual acuity, contrast sensitivity, retinal thickness, the a-wave and b-wave of scotopic and photic electroretinography (ERG), oscillatory potential (OP) amplitudes, 30-Hz flicker ERG amplitude, multifocal electroretinography (mfERG) amplitudes, electrooculography (EOG) Arden ratio, and the VEP amplitude from central or peripheral visual field and increase the latencies of ERG a-wave and b-wave [1–11]. Reduction in b-wave amplitudes and 30 Hz flicker amplitudes are usually the first signs of toxicity. There is also evidence that light exposure in mice administered VGB produces the taurine deficiency, causing the retinal degeneration [2, 12] Dysregulated expression of transcripts in the mTOR pathway, GABA/glutamate transporters $GABA_{A/B}$ receptors, and metabotropic glutamate receptors 1/6 in the eye are associated with the reduction of VEP amplitude after VGB administration [13].

Complex full-field electroretinography (ffERG) changes can be found in children placed on vigabatrin for intractable epilepsy (Fig. 9.1). However,

M. Yu (✉)
Department of Ophthalmology, University Hospitals Eye Institute, Cleveland, OH, USA
e-mail: minzhong.yu@uhhospitals.org

A. Senatore · A. Iannaccone · W. J. Kheir
Center for Retinal Degenerations and Ophthalmic Genetic Diseases, Duke University School of Medicine, Duke Eye Center, Department of Ophthalmology, Durham, NC, USA

D. Creel
Moran Eye Center, University of Utah School of Medicine, Salt Lake City, UT, USA

M. Yu et al. (eds.), *Handbook of Clinical Electrophysiology of Vision*,
https://doi.org/10.1007/978-3-030-30417-1_9

importantly, some ffERG changes can precede exposure to vigabatrin, underscoring that these patients may have preexisting changes in their retinal function related to their underlying condition. Thus, to correctly assess vigabatrin-mediated toxicity, obtaining baseline ffERGs is essential. With this caveat in mind, toxic effects during exposure to vigabatrin onto retinal function can affect both receptoral and post-receptoral responses and can vary from patient to patient (Fig. 9.1).

Hydroxychloroquine (Plaquenil) and Chloroquine

Hydroxychloroquine (HCQ) and chloroquine (CQ) are antimalarial agents commonly used in the treatment of lupus erythematosus, Sjogren's disease, rheumatoid arthritis, and dermatological inflammations. HCQ is also used in the treatment of uncomplicated malaria caused by chloroquine-sensitive strains of *Plasmodium* species or in the prophylaxis in regions with chloroquine-resistant strains. HCQ inhibits movements of neutrophils and eosinophils and also impairs complement-dependent antigen–antibody reactions. Its antirheumatic properties are proposed to result from their interference with "antigen processing" in macrophages and other antigen-presenting cells [14].

Both HCQ and CQ belong to the quinolone family and have similar clinical indications and side effects, including retinal toxicity. Hobbs et al. first described the CQ-induced retinal toxicity in 1959 [15], and Shearer et al. first described the HCQ toxicity to retina in 1967 [16]. HCQ has much less retinal side effect so that CQ has almost been replaced by HCQ for the treatment of inflammatory diseases. The retinal damage induced by these medications is usually irreversible. In a portion of patients with long-term HCQ/CQ treatment, HCQ-/CQ-induced toxicity can be found in cornea and macula, possibly related to the factors of cumulated dosage, duration of use, age, concomitant use of certain medications (e.g., tamoxifen), and health status (e.g., renal disease). In the retina, it may cause bilateral Bull's eye pattern of maculopathy (Fig. 9.2a), parafoveal damage of retina in early stage, and extramacular damage of retina in later stage. It affects ganglion cell layer, photoreceptors, and retinal pigment epithelium (RPE) by the inhibition of protein synthesis [17, 18] and peroxidation of lipid [19]. The risk of HCQ-/CQ-induced maculopathy is associated to the daily dose and the duration of HCQ/CQ administration [20, 21]. More damage may occur after the initial damage with continued use of HCQ/CQ [22]. Therefore, early screening of HCQ/CQ toxicity in these patients is necessary.

In mfERG test, HCQ-/CQ-induced maculopathy shows paracentral decrease of mfERG amplitude and prolongation of latency (Fig. 9.2b) [21, 23–30]. The ffERGs can be abnormal in some patients [31]. In the late stages of HCQ-induced retinopathy, ffERGs and other tests can be used together for the differential diagnosis [32]. For early detection of HCQ/CQ toxicity, mfERG, spectral domain optical coherence tomography (SD-OCT) (Fig. 9.2c), and fundus autofluorescence (FAF) (Fig. 9.2d)

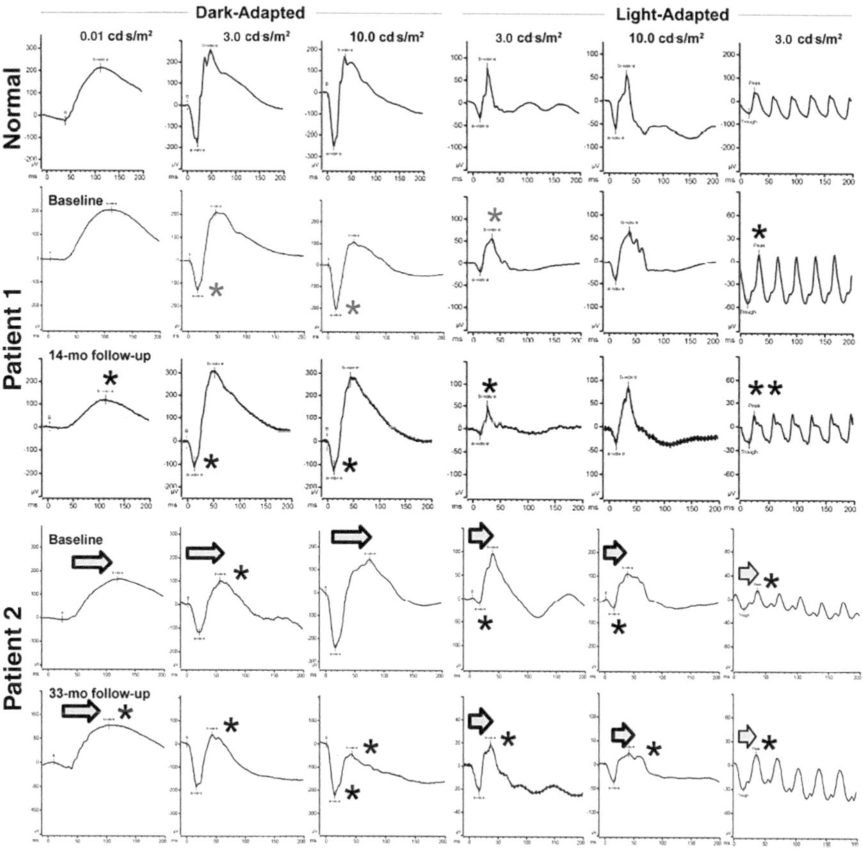

Fig. 9.1 Electroretinographic findings in vigabatrin-mediated retinal toxicity. Importantly, both patients exhibited *baseline* abnormalities, highlighting the importance of obtaining baseline measurements to ascertain potential retinal toxicities of this or any other medication. A normal set of ERGs, shown on the top row of the figure. *Patient 1.* This boy was tested first at baseline before the age of 2 years old, when he had first been placed on vigabatrin for intractable epilepsy linked to subdural hematomas and encephalopathy, and signs of toxicity emerged 14 months later. Dark- and light-adapted flash ERGs demonstrated at baseline mixed a-wave amplitude reduction, a mild electronegative shape of the 10.0 cd·s·m^{-2} response, and borderline low amplitudes for all cone-driven responses. After 14 months of follow-up, there was evidence of rod-driven b-wave attenuation, further reduction in mixed a-wave, lack of a-wave growth at 10 cd·s·m^{-2} (consistent with diminished photoreceptor sensitivity), and loss of cone-driven b-wave and flicker ERG amplitudes (note the different scale for flicker responses compared to normal). ON–OFF ERGs performed at follow-up (not shown) indicated also OFF > ON response compromise. Taken together, these findings suggested adverse vigabatrin-mediated effects at the photoreceptor and post-receptoral levels (on bipolar cells themselves or at the photoreceptor-to-bipolar-cell synapse). *Patient 2.* This girl was tested first at baseline before the age of 2 years old, when she had first been placed on vigabatrin for intractable epilepsy, and signs of toxicity emerged after 33 months. ERGs at baseline show delayed rod-driven ERGs, borderline low mixed dark-adapted a- and b-wave response amplitudes, delayed mixed b-waves, reduced photopic a-wave amplitudes, and attenuated and delayed flicker ERG responses. Compared to baseline, follow-up flash ERGs 33 months later demonstrate rod-driven and mixed b-wave amplitude loss (electronegative ERG), reduction in the photopic b-waves [arrows] with an artifactual a-wave increase (due to increased electronegativity in the light-adapted responses as well) attributable to ON>OFF post-receptoral response compromise (not shown). The flicker ERG was overall stable

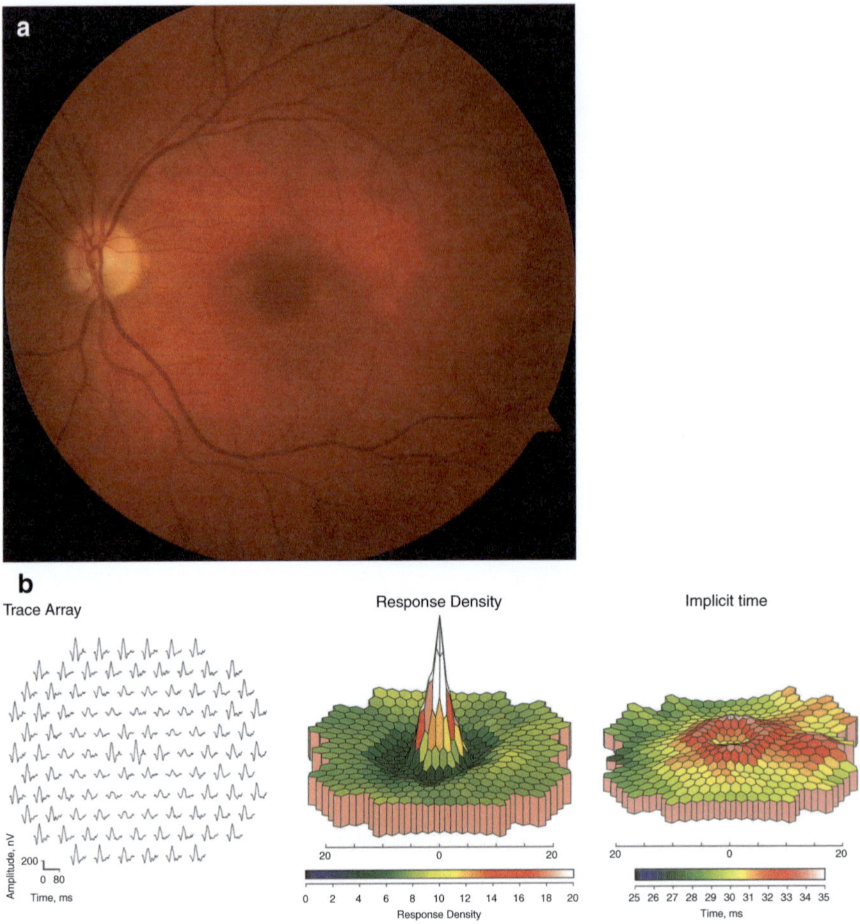

Fig. 9.2 Fundus imaging and mfERG in hydroxychloroquine retinopathy. (**a**) Fundus image of a hydroxychloroquine user showing Bull's eye pattern of maculopathy. (**b**) mfERG result of a hydroxychloroquine user demonstrates marked reduction of mfERG response density and prolongation of mfERG implicit time in the parafoveal region (Adapted from Figure 1 [21]). (**c**) SD-OCT of a hydroxychloroquine user that shows significant decrease of the thickness of outer retina in the parafoveal region. (**d**) Fundus autofluorescence displays higher signal in the parafoveal region

may have higher sensitivity than perimetry [24, 28, 33–37] and are part of the standard Plaquenil patient workup. In addition, mfERG can provide objective data of the retinal function across visual field [20].

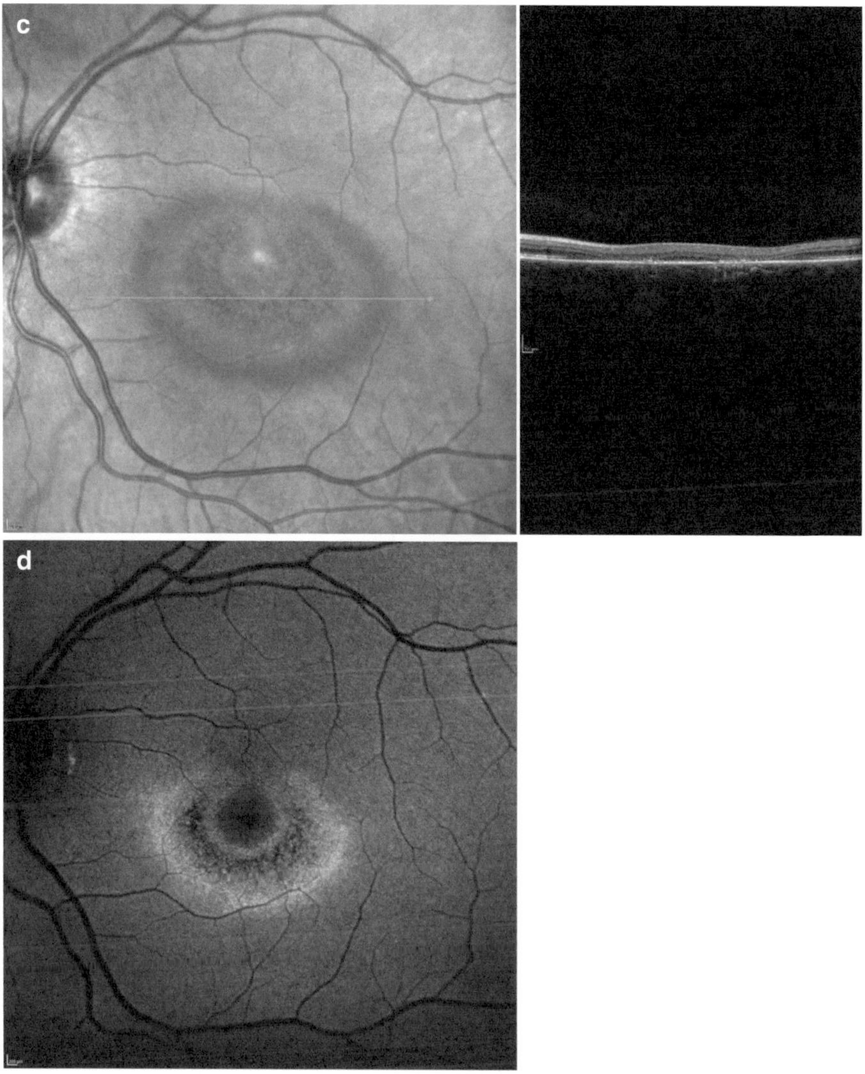

Fig. 9.2 (continued)

Antipsychotics

Chlorpromazine (CPZ) and thioridazine are members of the first-generation antipsychotic class, also known as typical antipsychotics. Their mechanism of action depends on the blockage of postsynaptic D2 receptors in the mesolimbic and mesocortical pathways. Typical antipsychotics may also block muscarinic, alpha 1-adrenergic, histamine, and 5-HT2 receptors.

CPZ administration can cause latency prolongation of the ffERG mixed response and cone response, decrease of rod and cone b-wave amplitude, and latency prolongation of the oscillatory potentials (OPs). However, CPZ is not known to change parameters of the fVEP, pERG, and the retinocortical times calculated from the difference of pERG and pVEP latencies [38].

Thioridazine (also known as Mellaril) is a dopaminergic antagonist (similar to CPZ) first used as antipsychotic with indication for the management of schizophrenic patients who fail to respond adequately to the treatment with other first-line antipsychotic drugs.

Because of its serious retinal and cardiac side effects (thioridazine may increase the risk of potentially fatal ventricular arrhythmias), its use is presently recommended only for selected cases. Probably due to an idiosyncratic response, retinal toxicity has also been reported for short-term and low-dose administrations [39]. However, long-term thioridazine treatments are usually associated with the highest risk of retinopathy.

The earliest signs of thioridazine retinal toxicity is RPE mottling of the macula and perimacular regions. Full-blown retinopathy is characterized by nummular (coin shaped) patches of RPE loss in both posterior pole and periphery, with a RP-like appearance [40, 41] (Fig. 9.3). Decreased visual acuity, night blindness, paracentral scotomas, and loss of peripheral vision can all occur in thioridazine retinopathy. In more advanced stages of the disease, there is definite electrophysiological (ffERG, EOG, and mfERG) evidence of diffuse retinal compromise (Figs. 9.4 and 9.5).

The mechanism of retinal damage is still unclear, but it is thought to be similar to chlorpromazine, accumulating mainly in pigmented cells including the uveal level [42]. This presumed mechanism may explain the extensive choriocapillaris loss and RPE dysfunction and damage co-contributing to the retinal damage. Direct dopaminergic mechanisms have also been suspected, but not proven.

Thioridazine discontinuation is the only treatment available. Regression has been described, but only in early-diagnosed cases, and further progression of retinal damage over several years after discontinuation can (more commonly) occur [43].

Cis-Platinum

Cis-platinum is a medicine used for the chemotherapy of bladder, testicular, ovarian, small and non-small cell lung, esophageal, breast, prostate, stomach, and cervical cancers. It is also used to treat mesothelioma, sarcomas, melanoma, multiple myeloma, neuroblastoma, and Hodgkin's and non-Hodgkin's lymphomas. Cis-platinum is an alkylating agent active during resting phase of the cell. Therefore, this drug is cell cycle nonspecific. The manifestation of electrophysiology of cis-platinum side effect includes reduction of ERG dark-adapted and light-adapted b-wave amplitudes, amplitude of OPs, amplitude of 30 Hz flicker ERG, and dark-adapted a-wave amplitude (Fig. 9.6) [44–46].

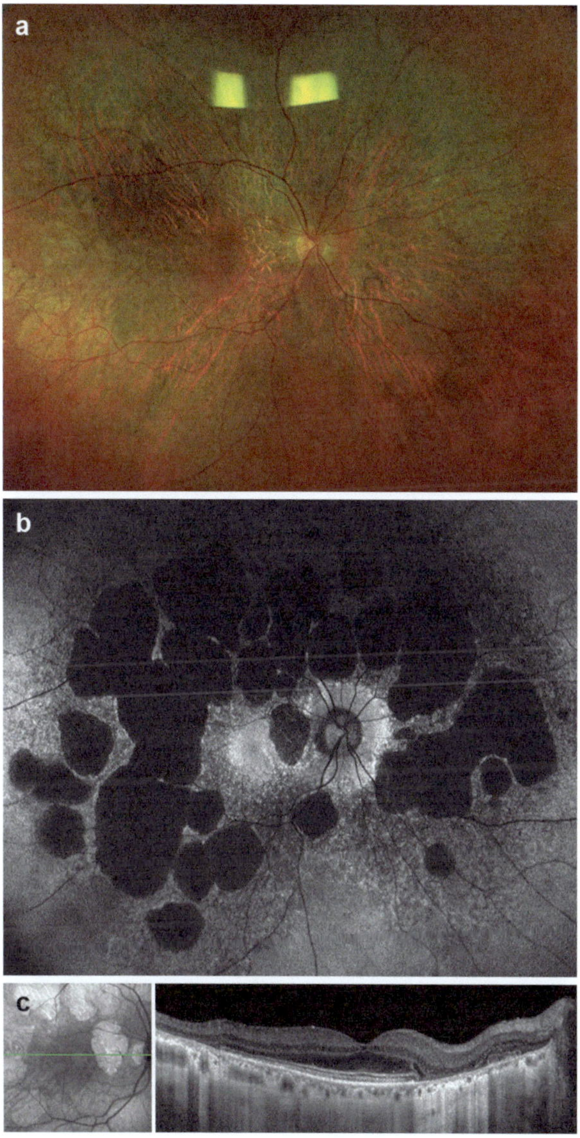

Fig. 9.3 Clinical and imaging findings in thioridazine retinopathy. (**a**) A large area of depigmentation across the entire posterior pole characterizes the fundus examination of this 70-year-old Caucasian woman with a previous longstanding history of treatment with thioridazine for a psychotic disorder. (**b**) Fundus autofluorescence imaging reveals much more clearly the nummular patches of hypo-autofluorescent lesions corresponding to the clinically visible atrophy of the RPE around the fovea and throughout the posterior pole, extending well beyond the arcades. There is central sparing but marked hyper-autofluorescence around the disc and in the foveal region. The nummular hypo-autofluorescent lesions are surrounded by a large halo of speckled hypo-autofluorescence indicative of further RPE suffering also in the periphery. (**c**) The OCT scan shows excellent central preservation of all layers, including the RPE, but sharp and discrete near-complete loss of RPE, EZ, ELM, and ONL just nasal to fovea and farther out in the temporal periphery. These clinical findings in the context of the thioridazine exposure are typical for thioridazine retinopathy

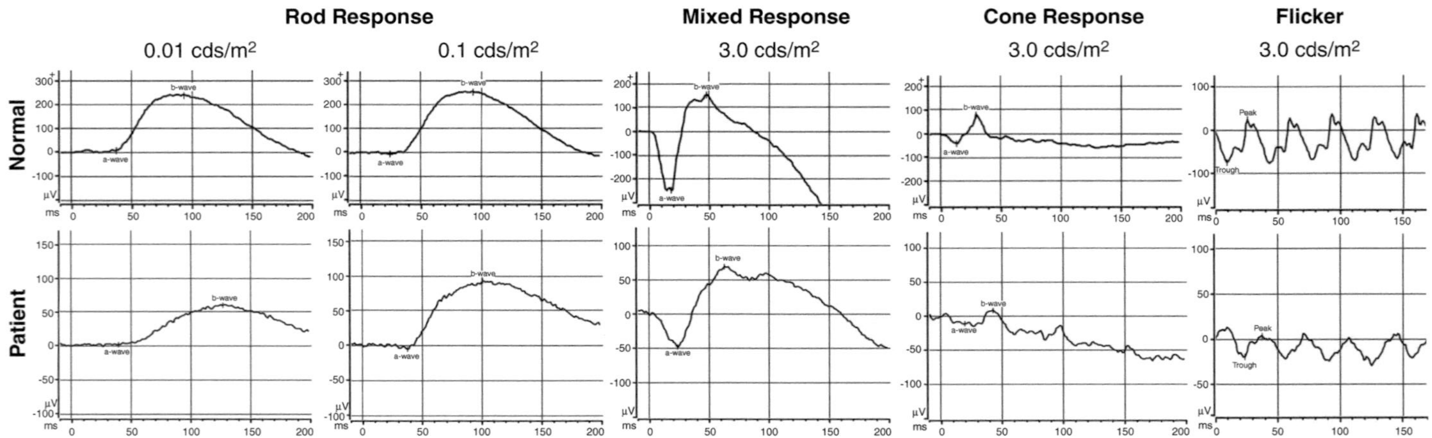

Fig. 9.4 Electroretinographic findings in thioridazine retinopathy. The ffERG of the patient illustrated in Fig. 9.3 reveals both rod and cone response compromise, in an approximately similar ratio, and marked response delays. Findings were symmetric in both eyes both from the ffERG and clinical perspective

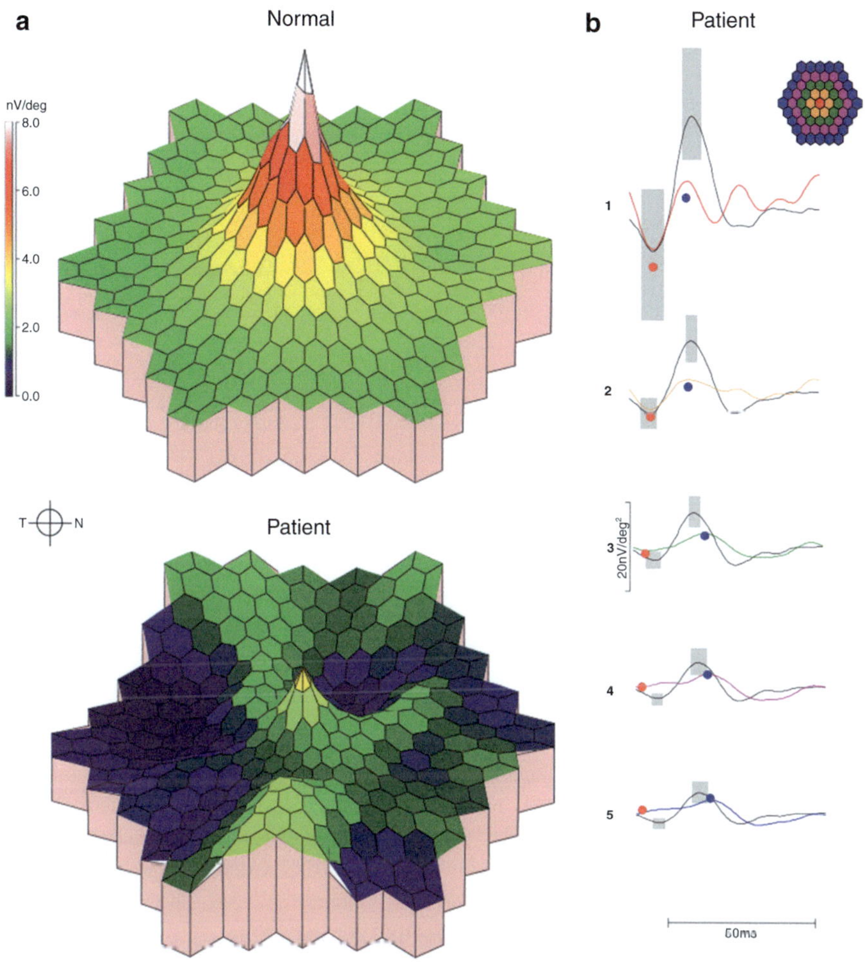

Fig. 9.5 Multifocal electroretinography findings in thioridazine retinopathy, compared with normal values. (**a**) 3D plots of the response densities. (**b**) Average responses in 5 rings of different eccentricities. The grey bars indicate the normal ranges of the N1 and P1 components with 95% confidence interval, respectively. The mfERG response of this patient (same as in Figs. 9.3 and 9.4) reveals patchy loss of the response density across the macular region, with relative preservation of the foveal peak, accounting for the excellent visual acuity. An EOG was also measured (not shown), and the Arden ratio was ≤1.35 in each eye, consistent with widespread RPE compromise, again consistent with the history of thioridazine exposure. Had this patient not had this history, a workup for a possible autosomal recessive bestrophinopathy would have been in order

Deferoxamine

Deferoxamine (DFO) is an IV detoxification agent used in both acute iron intoxication and chronic iron overload. It is also used for the detoxification of aluminum intoxication. DFO chelates iron from ferritin and hemosiderin and enhances its elimination. Both ocular and auditory disturbances have been reported with this agent,

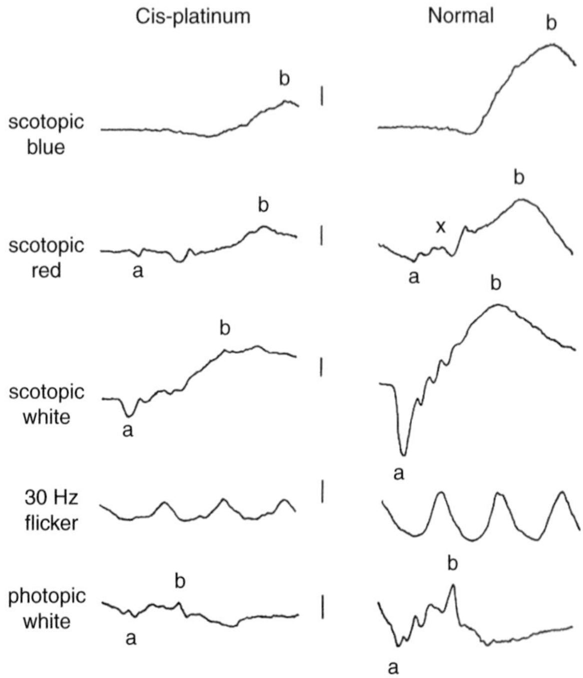

Fig. 9.6 Electroretinography findings in cis-platinum retinopathy. ffERG of cis-platinum retinopathy demonstrates the reduction of ffERG amplitudes in both rod and cone pathways

suggesting pigmentary retinopathy and optic neuropathy can be caused by DFO. The ocular symptoms of its side effects include vision loss, dyschromatopsia, nyctalopia, and scotomas that can be partially recovered after the cessation of DFO administration. Electrophysiology can be used to monitor this side effect seen in the reduction of ERG responses of rod and cone pathways (Fig. 9.7), EOG Arden ratio, mfERG amplitude in central field, and the prolongation of VEP latency [47–51]. The visual function may be recovered after the cessation of the DFO administration.

Digoxin

Digoxin is a drug in the cardiac glycoside family of medications for the treatment of some cardiac disorders, including heart failure, atrial fibrillation and atrial flutter. Its side effects include abnormal vision, irregular heartbeat, nausea, loss of appetite, confusion, and breast enlargement. The symptoms of the visual disorder include blurry vision, central scotomas, photopsia, abnormal color vision, and xanthopsia [52–55]. Studies show ERG or mfERG in the central field in some patients treated with digoxin is abnormal [56–59]. Some studies with animals also confirmed the side effect of digoxin on retinal cone and rod photoreceptors displayed as abnormal ERG results and other tests of visual function and morphology [60–63].

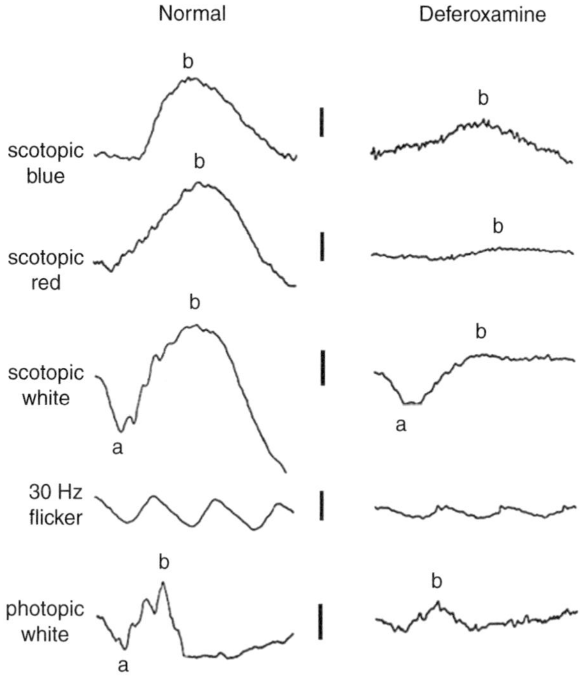

Fig. 9.7 Electroretinography findings in deferoxamine retinopathy. All responses of rod and cone pathways reduce significantly in deferoxamine retinopathy

Ethambutol

Ethambutol (EMB) is mainly used to treat tuberculosis. Its side effects include ethambutol-induced optic neuropathy and retinopathy. For early detection of ethambutol-induced retinopathy, mfERG can show the decrease of P1 amplitudes and/or prolonged P1 latencies in the central, bitemporal, peripheral, nasal, or the whole tested field even before the visual symptoms occur [64–69]. These papers suggest mfERG is probably suitable for early diagnosis of ethambutol-induced retinopathy.

Indomethacin

Indomethacin is a nonsteroidal, anti-inflammatory drug that can be used to treat fever, pain, and inflammation. Indomethacin is a reversible inhibitor of cyclooxygenase 1 and 2, leading to decreased production of prostaglandins. Indomethacin can cause retinopathy reflected in significant reduction of rod and cone responses of ERG and the Arden ratio of EOG [70].

Isotretinoin

Isotretinoin is a medication primarily used to treat severe acne unresponsive to conventional therapy and occasionally used in the treatment and prevention of cancer. Isotrentinoin's exact mechanism of action is unknown. However, isotretinoin may induce apoptosis. Due to its high teratogenicity, isotretinoin is distributed under a risk evaluation and mitigation strategy (REMS) program called iPledge, intended to prevent fetal exposure to the drug. The ocular side effect of isotretinoin can be manifested by the reduction of scotopic ERG amplitudes and Arden ratio of EOG [71–73].

Ocular Siderosis

Ocular siderosis (OS) is caused by retained iron-containing intraocular foreign body after a penetrating ocular injury (Fig. 9.8a) that can cause significant vision loss. Pigmentary retinopathy and cataract are common in OS. The degree of retinal damage and recovery is affected by many factors, such as location and size of the foreign body, which can be monitored by the decrease and improvement of ffERG amplitudes (Fig. 9.8b) [74–79], mfERG latency [80], the relationship of a-wave and b-wave amplitudes [81], or the Arden ratio of EOG [82].

Phosphodiesterase (PDE) Type 5 Inhibitors (Erectile Dysfunction Medication)

Phosphodiesterase (PDE) type 5 inhibitors, such as sildenafil, tadalafil, vardenafil, udenafil, and avanafil, are commonly used in erectile dysfunction and pulmonary hypertension. PDE5 inhibition leads to decreased degradation of cGMP leading to vasodilation. Sexual stimulation leads to a physiological release of nitric oxide activating guanylate cyclase and production of cGMP. Increased levels of cGMP from the medication leads to smooth muscle relaxation and increased blood flow to the penis. However, these drugs may cause transient retinal disorders or optic neuropathy [83, 84].

The side effect in patients includes abnormal color vision, blurred vision, photophobia, ocular pain, and conjunctival hyperemia [85]. Decrease of amplitude and prolongation of latency in ffERG and mfERG were observed in some subjects or animals without preexisting ocular disorders after sildenafil administration [86–91]. Decrease of response in retinal ganglion cells in rat after sildenafil administration was also observed [92]. However, sildenafil increased the a-wave and b-wave amplitude in a rat model of neonatal hypoxia-ischemia [93] and the b-wave amplitude in a mouse carrier of retinitis pigmentosa [88]. In addition, both sildenafil and tadalafil may cause central serous chorioretinopathy [94, 95] and damage photoreceptors

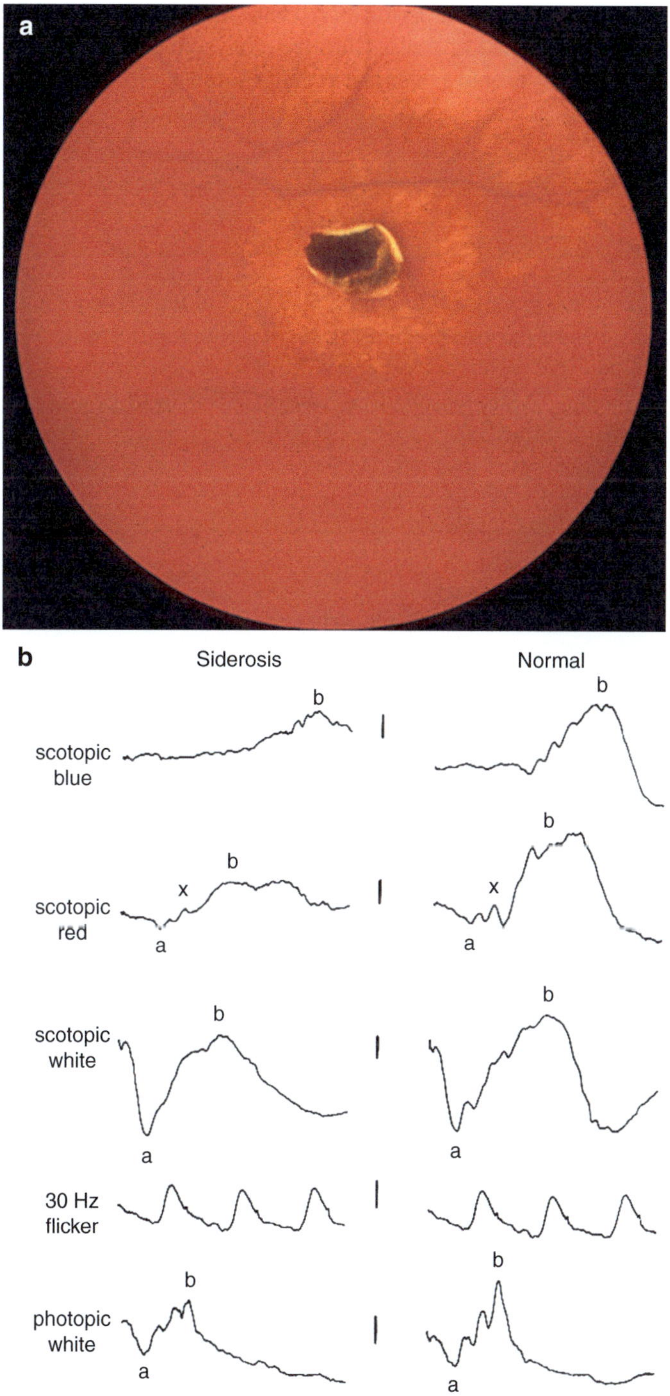

Fig. 9.8 A case of fundus and ffERG findings in ocular siderosis. (**a**) A foreign body in the retina. (**b**) ffERG shows that the rod responses are reduced while cone responses are relatively intact

[96]. Sildenafil, tadalafil, and udenafil may cause optic neuropathy [91, 97–108]. Therefore, it is complicated to explain the electrophysiologic results of the users of PDE5 inhibitor for the diagnosis of other ocular disorders.

Quinine

Quinine is used in the treatment of uncomplicated malaria where resistance to chloroquine has been documented. Quinine is also used in the prevention of nocturnal leg cramps. The exact mechanism of antimalarial activity is not known. However, it is thought quinine can inhibit nucleic acid synthesis, protein synthesis, and glycolysis. The amplitudes of ffERG, mfERG, VEP, and EOG can be reduced after the administration of quinine [109–114].

References

1. Hebert-Lalonde N, et al. Electrophysiological evidences of visual field alterations in children exposed to vigabatrin early in life. Pediatr Neurol. 2016;59:47–53.
2. Tao Y, et al. The vigabatrin induced retinal toxicity is associated with photopic exposure and taurine deficiency: an in vivo study. Cell Physiol Biochem. 2016;40(5):831–46.
3. Kjellstrom U, Andreasson S, Ponjavic V. Attenuation of the retinal nerve fibre layer and reduced retinal function assessed by optical coherence tomography and full-field electroretinography in patients exposed to vigabatrin medication. Acta Ophthalmol. 2014;92(2):149–57.
4. Ruether K, et al. Electrophysiologic evaluation of a patient with peripheral visual field contraction associated with vigabatrin. Arch Ophthalmol. 1998;116(6):817–9.
5. Harding GF, et al. Electro-oculography, electroretinography, visual evoked potentials, and multifocal electroretinography in patients with vigabatrin-attributed visual field constriction. Epilepsia. 2000;41(11):1420–31.
6. Eke T, Talbot JF, Lawden MC. Severe persistent visual field constriction associated with vigabatrin. BMJ. 1997;314(7075):180–1.
7. Gross-Tsur V, et al. Visual impairment in children with epilepsy treated with vigabatrin. Ann Neurol. 2000;48(1):60–4.
8. Daneshvar H, et al. Symptomatic and asymptomatic visual loss in patients taking vigabatrin. Ophthalmology. 1999;106(9):1792–8.
9. Schroeder CE, et al. Effects of high-dose gamma-vinyl GABA (vigabatrin) administration on visual and somatosensory evoked potentials in dogs. Epilepsia. 1992;33 Suppl 5:S13–25.
10. Cosi V, et al. Effects of vigabatrin on evoked potentials in epileptic patients. Br J Clin Pharmacol. 1989;27(Suppl 1):61S–8S.
11. Ponjavic V, Andreasson S. Multifocal ERG and full-field ERG in patients on long-term vigabatrin medication. Doc Ophthalmol. 2001;102(1):63–72.
12. Yang J, et al. Vigabatrin-induced retinal toxicity is partially mediated by signaling in rod and cone photoreceptors. PLoS One. 2012;7(8):e43889.
13. Vogel KR, et al. mTOR inhibition mitigates molecular and biochemical alterations of vigabatrin-induced visual field toxicity in mice. Pediatr Neurol. 2017;66:44–52 e1.
14. Fox RI. Mechanism of action of hydroxychloroquine as an antirheumatic drug. Semin Arthritis Rheum. 1993;23(2 Suppl 1):82–91.

15. Hobbs HE, Sorsby A, Freedman A. Retinopathy following chloroquine therapy. Lancet. 1959;2(7101):478–80.
16. Shearer RV, Dubois EL. Ocular changes induced by long-term hydroxychloroquine (plaquenil) therapy. Am J Ophthalmol. 1967;64(2):245–52.
17. Ivanina TA, et al. A study of the mechanisms of chloroquine retinopathy. II. Chloroquine effect on protein synthesis of retina. Ophthalmic Res. 1989;21(3):272–7.
18. Gonasun LM, Potts AM. In vitro inhibition of protein synthesis in the retinal pigment epithelium by chloroquine. Investig Ophthalmol. 1974;13(2):107–15.
19. Ivanina TA, et al. A study of the mechanisms of chloroquine retinopathy. I. Chloroquine effect on lipid peroxidation of retina. Ophthalmic Res. 1989;21(3):216–20.
20. Marmor MF, et al. Recommendations on screening for chloroquine and hydroxychloroquine retinopathy (2016 revision). Ophthalmology. 2016;123(6):1386–94.
21. Maturi RK, Yu M, Weleber RG. Multifocal electroretinographic evaluation of long-term hydroxychloroquine users. Arch Ophthalmol. 2004;122(7):973–81.
22. Pasadhika S, Fishman GA. Effects of chronic exposure to hydroxychloroquine or chloroquine on inner retinal structures. Eye (Lond). 2010;24(2):340–6.
23. So SC, et al. Evaluation of hydroxychloroquine retinopathy with multifocal electroretinography. Ophthalmic Surg Lasers Imaging. 2003;34(3):251–8.
24. Michaelides M, et al. Retinal toxicity associated with hydroxychloroquine and chloroquine: risk factors, screening, and progression despite cessation of therapy. Arch Ophthalmol. 2011;129(1):30–9.
25. Kellner U, Kraus H, Foerster MH. Multifocal ERG in chloroquine retinopathy: regional variance of retinal dysfunction. Graefes Arch Clin Exp Ophthalmol. 2000;238(1):94–7.
26. Lai TY, et al. Multifocal electroretinographic changes in patients receiving hydroxychloroquine therapy. Am J Ophthalmol. 2005;140(5):794–807.
27. Chang WH, et al. A novel method for screening the multifocal electroretonogram in patients using hydroxychloroquine. Retina. 2008;28(10):1478–86.
28. Lyons JS, Severns ML. Using multifocal ERG ring ratios to detect and follow Plaquenil retinal toxicity: a review: review of mfERG ring ratios in Plaquenil toxicity. Doc Ophthalmol. 2009;118(1):29–36.
29. Tsang AC, et al. Hydroxychloroquine and chloroquine retinopathy: a systematic review evaluating the multifocal electroretinogram as a screening test. Ophthalmology. 2015;122(6):1239–1251 e4.
30. Moschos MN, et al. Assessing hydroxychloroquine toxicity by the multifocal ERG. Doc Ophthalmol. 2004;108(1):47–53.
31. Tzekov RT, Serrato A, Marmor MF. ERG findings in patients using hydroxychloroquine. Doc Ophthalmol. 2004;108(1):87–97.
32. Nair AA, Marmor MF. ERG and other discriminators between advanced hydroxychloroquine retinopathy and retinitis pigmentosa. Doc Ophthalmol. 2017;134(3):175–83.
33. Marmor MF, et al. Revised recommendations on screening for chloroquine and hydroxychloroquine retinopathy. Ophthalmology. 2011;118(2):415–22.
34. Young B, Eggenberger E, Kaufman D. Current electrophysiology in ophthalmology: a review. Curr Opin Ophthalmol. 2012;23(6):497–505.
35. Farrell DF. Retinal toxicity to antimalarial drugs: chloroquine and hydroxychloroquine: a neurophysiologic study. Clin Ophthalmol. 2012;6:377–83.
36. Nebbioso M, et al. Retina in rheumatic diseases: standard full field and multifocal electroretinography in hydroxychloroquine retinal dysfunction. Clin Exp Optom. 2011;94(3):276–83.
37. Nebbioso M, Grenga R, Karavitis P. Early detection of macular changes with multifocal ERG in patients on antimalarial drug therapy. J Ocul Pharmacol Ther. 2009;25(3):249–58.
38. Bartel P, et al. Effects of chlorpromazine on pattern and flash ERGs and VEPs compared to oxazepam and to placebo in normal subjects. Electroencephalogr Clin Neurophysiol. 1990;77(5):330–9.

39. Neves MS, Jordon K, Dragt H. Extensive chorioretinopathy associated with very low dose thioridazine. Eye (Lond). 1990;4(Pt 5):767–70.
40. Hagopian V, Stratton DB, Busiek RD. Five cases of pigmentary retinopathy associated with thioridazine administration. Am J Psychiatry. 1966;123(1):97–100.
41. Kozy D, Doft BH, Lipkowitz J. Nummular thioridazine retinopathy. Retina. 1984;4(4):253–6.
42. Persad S, et al. Phototoxicity of chlorpromazine on retinal pigment epithelial cells. Curr Eye Res. 1988;7(1):1–9.
43. Richa S, Yazbek JC. Ocular adverse effects of common psychotropic agents: a review. CNS Drugs. 2010;24(6):501–26.
44. Katz BJ, et al. Persistent severe visual and electroretinographic abnormalities after intravenous Cisplatin therapy. J Neuroophthalmol. 2003;23(2):132–5.
45. Marmor MF. Negative-type electroretinogram from cisplatin toxicity. Doc Ophthalmol. 1993;84(3):237–46.
46. Wilding G, et al. Retinal toxicity after high-dose cisplatin therapy. J Clin Oncol. 1985;3(12):1683–9.
47. Simon S, et al. Desferrioxamine-related ocular toxicity: a case report. Indian J Ophthalmol. 2012;60(4):315–7.
48. Kaplinsky C, et al. Deferoxamine (Desferal)-induced ocular toxicity. Pediatr Hematol Oncol. 1988;5(4):293–7.
49. Haimovici R, et al. The expanded clinical spectrum of deferoxamine retinopathy. Ophthalmology. 2002;109(1):164–71.
50. Schmidt D, Finke J. Bull's-eye maculopathy with deferoxamine treatment. Klin Monatsbl Augenheilkd. 2004;221(3):204–9.
51. Kertes PJ, Lee TK, Coupland SG. The utility of multifocal electroretinography in monitoring drug toxicity: deferoxamine retinopathy. Can J Ophthalmol. 2004;39(6):656–61.
52. Lely AH, van Enter CH. Large-scale digitoxin intoxication. Br Med J. 1970;3(5725):737–40.
53. Butler VP Jr, et al. Digitalis-induced visual disturbances with therapeutic serum digitalis concentrations. Ann Intern Med. 1995;123(9):676–80.
54. Robertson DM, Hollenhorst RW, Callahan JA. Ocular manifestations of digitalis toxicity. Discussion and report of three cases of central scotomas. Arch Ophthalmol. 1966;76(5):640–5.
55. Lawrenson JG, et al. Acquired colour vision deficiency in patients receiving digoxin maintenance therapy. Br J Ophthalmol. 2002;86(11):1259–61.
56. Renard D, et al. Spectrum of digoxin-induced ocular toxicity: a case report and literature review. BMC Res Notes. 2015;8:368.
57. Piltz JR, et al. Digoxin toxicity. Recognizing the varied visual presentations. J Clin Neuroophthalmol. 1993;13(4):275–80.
58. Weleber RG, Shults WT. Digoxin retinal toxicity. Clinical and electrophysiological evaluation of a cone dysfunction syndrome. Arch Ophthalmol. 1981;99(9):1568–72.
59. Madreperla SA, Johnson MA, Nakatani K. Electrophysiologic and electroretinographic evidence for photoreceptor dysfunction as a toxic effect of digoxin. Arch Ophthalmol. 1994;112(6):807–12.
60. Landfried B, et al. Digoxin-induced retinal degeneration depends on rhodopsin. Cell Death Dis. 2017;8(3):e2670.
61. Kinoshita J, et al. Digoxin-induced reversible dysfunction of the cone photoreceptors in monkeys. Invest Ophthalmol Vis Sci. 2014;55(2):881–92.
62. Maehara S, et al. Detection of cone dysfunction induced by digoxin in dogs by multicolor electroretinography. Vet Ophthalmol. 2005;8(6):407–13.
63. Deeti S, O'Farrell S, Kennedy BN. Early safety assessment of human oculotoxic drugs using the zebrafish visualmotor response. J Pharmacol Toxicol Methods. 2014;69(1):1–8.
64. Kandel H, et al. Visual function in patients on ethambutol therapy for tuberculosis. J Ocul Pharmacol Ther. 2012;28(2):174–8.
65. Lai TY, et al. Multifocal electroretinogram demonstrated macular toxicity associated with ethambutol related optic neuropathy. Br J Ophthalmol. 2005;89(6):774–5.

66. Behbehani RS, et al. Multifocal ERG in ethambutol associated visual loss. Br J Ophthalmol. 2005;89(8):976–82.
67. Kardon RH, Morrisey MC, Lee AG. Abnormal multifocal electroretinogram (mfERG) in ethambutol toxicity. Semin Ophthalmol. 2006;21(4):215–22.
68. Lai TY, et al. Multifocal electroretinography changes in patients on ethambutol therapy. Eye (Lond). 2009;23(8):1707–13.
69. Liu Y, et al. Multifocal electroretinographic abnormalities in ethambutol-induced visual loss. J Neuroophthalmol. 2008;28(4):278–82.
70. Graham CM, Blach RK. Indomethacin retinopathy: case report and review. Br J Ophthalmol. 1988;72(6):434–8.
71. Madke B, Prasad K, Kar S. Isotretinoin-induced night blindness. Indian J Dermatol. 2015;60(4):424.
72. Weleber RG, et al. Abnormal retinal function associated with isotretinoin therapy for acne. Arch Ophthalmol. 1986;104(6):831–7.
73. Mollan SP, et al. Does use of isotretinoin rule out a career in flying? Br J Ophthalmol. 2006;90(8):957–9.
74. Kannan NB, et al. Management of ocular siderosis: visual outcome and electroretinographic changes. J Ophthalmol. 2016;2016:7272465.
75. Mumcuoglu T, et al. An animal model (guinea pig) of ocular siderosis: histopathology, pharmacology, and electrophysiology. Curr Eye Res. 2015;40(3):314–20.
76. Faure C, et al. Functional and high resolution retinal imaging assessment in a case of ocular siderosis. Doc Ophthalmol. 2014;128(1):69–75.
77. Kuhn F, et al. Improvement of siderotic ERG. Eur J Ophthalmol. 1992;2(1):44–5.
78. Declercq SS. Desferrioxamine in ocular siderosis: a long-term electrophysiological evaluation. Br J Ophthalmol. 1980;64(8):626–9.
79. Declercq SS, Meredith PC, Rosenthal AR. Experimental siderosis in the rabbit: correlation between electroretinography and histopathology. Arch Ophthalmol. 1977;95(6):1051–8.
80. Gupta S, et al. Sensitivity of multifocal electroretinography (mfERG) in detecting siderosis. Can J Ophthalmol. 2015;50(6):485–90.
81. Schechner R, et al. A long term follow up of ocular siderosis: quantitative assessment of the electroretinogram. Doc Ophthalmol. 1990;76(3):231–40.
82. Schocket SS, Lakhanpal V, Varma SD. Siderosis from a retained intraocular stone. Retina. 1981;1(3):201–7.
83. Eltony SA, Abdelhameed SY. Effect of chronic administration of sildenafil citrate (Viagra) on the histology of the retina and optic nerve of adult male rat. Tissue Cell. 2017;49(2. Pt B):323–35.
84. Coca MN, et al. Bilateral posterior ischemic optic neuropathy associated with the use of Sildenafil for pulmonary hypertension. Can J Ophthalmol. 2016;51(3):e96–9.
85. Moschos MM, Nitoda E. Pathophysiology of visual disorders induced by phosphodiesterase inhibitors in the treatment of erectile dysfunction. Drug Des Devel Ther. 2016;8:3407–13.
86. Luu JK, et al. Acute effects of sildenafil on the electroretinogram and multifocal electroretinogram. Am J Ophthalmol. 2001;132(3):388–94.
87. Kinoshita J, et al. Sildenafil-induced reversible impairment of rod and cone phototransduction in monkeys. Invest Ophthalmol Vis Sci. 2015;56(1):664–73.
88. Nivison-Smith L, et al. Sildenafil alters retinal function in mouse carriers of retinitis pigmentosa. Exp Eye Res. 2014;128:43–56.
89. Jagle H, et al. Dose-dependency and time-course of electrophysiologic short-term effects of VIAGRA: a case study. Doc Ophthalmol. 2005;110(2–3):247–54.
90. Luke M, et al. Effects of phosphodiesterase type 5 inhibitor sildenafil on retinal function in isolated superfused retina. J Ocul Pharmacol Ther. 2005;21(4):305–14.
91. McKoy JM, et al. Sildenafil- and tadalafil-associated optic neuropathy: implications for men after prostate cancer treatment. Commun Oncol. 2009;6(2):78–80.

92. Martins J, et al. Sildenafil acutely decreases visual responses in ON and OFF retinal ganglion cells. Invest Ophthalmol Vis Sci. 2015;56(4):2639–48.
93. Jung S, et al. Sildenafil improves functional and structural outcome of retinal injury following term neonatal hypoxia-ischemia. Invest Ophthalmol Vis Sci. 2016;57(10):4306–14.
94. Roy R, et al. Central serous chorioretinopathy following oral tadalafil intake. Clin Exp Optom. 2014;97(5):473–4.
95. Fraunfelder FW, Fraunfelder FT. Central serous chorioretinopathy associated with sildenafil. Retina. 2008;28(4):606–9.
96. Coscas F, et al. Optical coherence tomography in tadalafil-associated retinal toxicity. Eur J Ophthalmol. 2012;22(5):853–6.
97. Peter NM, Singh MV, Fox PD. Tadalafil-associated anterior ischaemic optic neuropathy. Eye (Lond). 2005;19(6):715–7.
98. Egan R, Pomeranz H. Sildenafil (Viagra) associated anterior ischemic optic neuropathy. Arch Ophthalmol. 2000;118(2):291–2.
99. Cunningham AV, Smith KH. Anterior ischemic optic neuropathy associated with viagra. J Neuroophthalmol. 2001;21(1):22–5.
100. Boshier A, Pambakian N, Shakir SA. A case of nonarteritic ischemic optic neuropathy (NAION) in a male patient taking sildenafil. Int J Clin Pharmacol Ther. 2002;40(9):422–3.
101. Dheer S, Rekhi GS, Merlyn S. Sildenafil associated anterior ischaemic optic neuropathy. J Assoc Physicians India. 2002;50:265.
102. Pomeranz HD, et al. Sildenafil-associated nonarteritic anterior ischemic optic neuropathy. Ophthalmology. 2002;109(3):584–7.
103. Gruhn N, Fledelius HC. Unilateral optic neuropathy associated with sildenafil intake. Acta Ophthalmol Scand. 2005;83(1):131–2.
104. Pomeranz HD, Bhavsar AR. Nonarteritic ischemic optic neuropathy developing soon after use of sildenafil (viagra): a report of seven new cases. J Neuroophthalmol. 2005;25(1):9–13.
105. Escaravage GK Jr, Wright JD Jr, Givre SJ. Tadalafil associated with anterior ischemic optic neuropathy. Arch Ophthalmol. 2005;123(3):399–400.
106. Bollinger K, Lee MS. Recurrent visual field defect and ischemic optic neuropathy associated with tadalafil rechallenge. Arch Ophthalmol. 2005;123(3):400–1.
107. Egan RA, Fraunfelder FW. Viagra and anterior ischemic optic neuropathy. Arch Ophthalmol. 2005;123(5):709–10.
108. Kim IG, Kim DY. Anterior ischemic optic neuropathy associated with udenafil. Korean J Ophthalmol. 2012;26(3):235–8.
109. Moloney JB, Hillery M, Fenton M. Two year electrophysiology follow-up in quinine amblyopia. A case report. Acta Ophthalmol. 1987;65(6):731–4.
110. Su D, et al. Quinine toxicity: multimodal retinal imaging and electroretinography findings. Retin Cases Brief Rep. 2017;11 Suppl 1:S102–6.
111. Gangitano JL, Keltner JL. Abnormalities of the pupil and visual-evoked potential in quinine amblyopia. Am J Ophthalmol. 1980;89(3):425–30.
112. Sato S. Toxic effects on the visual system of diaminodiphenoxybutane, quinine, and ethambutol in conscious dogs. Fundam Appl Toxicol. 1985;5(4):777–84.
113. Saeed MU, et al. Relatively spared central multifocal electroretinogram responses in acute quinine toxicity. BMJ Case Rep. 2011;2011.
114. Verdon W. Clinical electrophysiology in quinine induced retinal toxicity. Optom Vis Sci. 2008;85(1):17–26.

Chapter 10
Characteristics of Visual Electrophysiology in the Diseases of Optic Nerve or Visual Pathway

Minzhong Yu and Donnell Creel

Multiple Sclerosis

Multiple sclerosis (MS) is a chronic disease that usually causes demyelination of the axons of nerve cells and brain cells due to autoimmune disease. The optic nerve is composed of retinal ganglion cell axons and glial cells. The axons are covered with myelin, which is usually disrupted in MS.

The myelin sheath insulates the axon of nerve cell between the nodes, which increases conduction velocity of signals transferred along the axon. Demyelination disrupts the myelin sheath and reduces the velocity of signal conduction.

In most patients with MS, the pattern reversal VEP latencies are prolonged, and the amplitudes are reduced in later stages of MS (Fig. 10.1) [1]. The b-wave of the photopic ERG can show attenuation of amplitude and prolongation of implicit time. The amplitude of flicker ERG can be reduced [2].

Ischemic Optic Neuropathy

Ischemic optic neuropathy (ION) is a group of disorders of the optic nerve due to the restricted blood supply [3]. The blood supply to the optic disc in the anterior portion of optic nerve is provided by the pial plexus and the central retinal artery, which do not provide blood supply to the rest of the optic nerve [4]. The blood supply of the retrobulbar portion of optic nerve is only provided by the circumfer-

M. Yu (✉)
Department of Ophthalmology, University Hospitals Eye Institute, Cleveland, OH, USA
e-mail: minzhong.yu@uhhospitals.org

D. Creel
Moran Eye Center, University of Utah School of Medicine, Salt Lake City, UT, USA

M. Yu et al. (eds.), *Handbook of Clinical Electrophysiology of Vision*,
https://doi.org/10.1007/978-3-030-30417-1_10

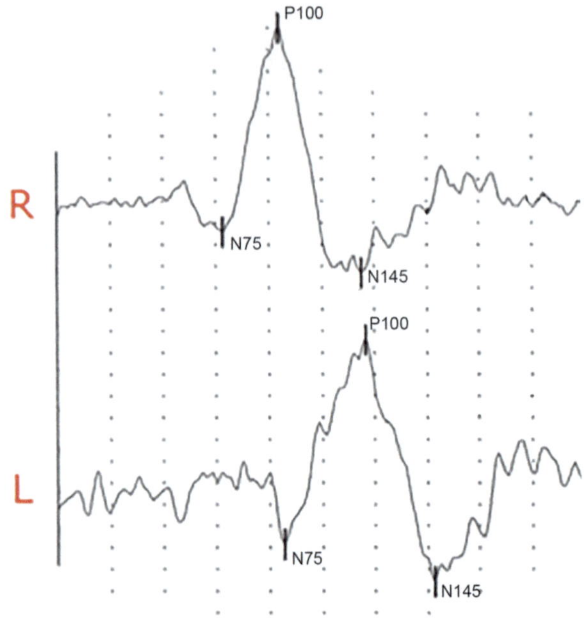

Fig. 10.1 Pattern reversal VEPs recorded from an adult female with MS and unilateral optic neuritis. N75, P100, and N145 waves are labeled. VEP traces of right and left eye stimulation are displayed. The vertical tick marks are 1 μV intervals. The VEP peak times of right eye are in normal range. The VEP peak times of left eye are significantly slow. Time epoch for VEP is 250 msec

ential pial capillary plexus derived from collateral distal branches of the ophthalmic artery that surrounds it [5]. ION includes anterior ischemic optic neuropathy (AION) and posterior ischemic optic neuropathy (PION), according to the location of ischemia. However, PION may have disruption in the anterior portion of optic nerve as well in addition to the posterior portion. The most common causes of ION include hypertension, high cholesterol level, diabetes, and anatomically crowded disc. Some surgeries may induce ION as well. In cardiac surgeries, AION occurs more frequently than PION [6] while PION is more common in spinal procedures and radical neck dissections, especially in prolonged spinal surgeries in prone position [7–9].

AION is distinguished from PION by the fact that AION occurs spontaneously and unilaterally in patients with predisposing anatomic or cardiovascular risk factors. Eyes with AION show optic disc edema acutely, while eyes with PION manifest a normal appearance of fundus. The diagnosis of PION is more difficult because the optic nerves initially appear normal.

In AION, OPs, pERG, and VEP are abnormal. The decrease of VEP amplitude and the normal to mild prolongation of VEP latency are more common in AION [10, 11]. PION shows the similar characteristics of electrophysiology as AION so that one cannot use electrophysiology to distinguish between AION and PION [12, 13]. However, the latency prolongation in ION is much less than those in optic neuritis and compressive optic neuropathy, which can assist the differentiation of these diseases [14].

Optic Neuritis

Optic neuritis (ON) is an inflammation in the optic nerve. The symptoms include pain, temporary vision loss, flashes of light, and visual field loss in the affected eye. ON can also happen during the development of multiple sclerosis or with other infections or immune diseases. Symptoms of optic neuritis may be the first indication of multiple sclerosis.

In ON, flash ERGs can be normal; pERG amplitude decreases significantly; pattern reversal VEP amplitude decreases as disease progresses, and the P100 latency increases significantly (Fig. 10.2) [15].

Optic Nerve Hypoplasia

Optic nerve hypoplasia (ONH), also known as "De Morsier syndrome" or "septo-optic dysplasia," is a congenital disease due to the underdevelopment of the optic nerve. The optic disc is smaller than normal in ONH. This disease is often associated with developmental delay, brain or pituitary malformations or other endocrinopathies, and related to the lack of growth hormone. ONH patients usually have nystagmus with visual acuity from no light perception to normal.

ONH shows significant decrease of amplitude in pERG and VEP, while flash ERG a- and b-wave can be abnormal or normal [16–18].

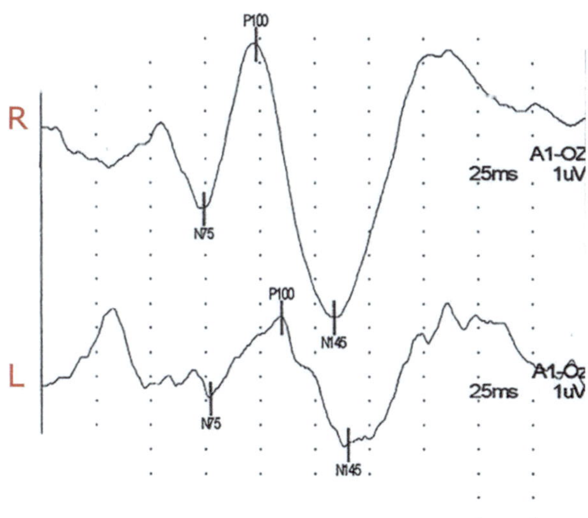

Fig. 10.2 Pattern reversal VEPs recorded from a patient with later-stage optic neuritis showing both the marked prolongation of VEP latency and reduction of VEP amplitude in the left optic nerve. Time epoch for VEP is 250 msec

Traumatic Optic Neuropathy

Traumatic optic neuropathy (TON) is an acute injury of the optic nerve caused by trauma. The axons of optic nerve may be disrupted that causes partial or complete visual loss. An indirect injury to the optic nerve occurs from the forces transmitted from blunt eye trauma to the optic canal. In contrast, a direct TON is caused by an anatomical disruption of the optic nerve fibers from penetrating orbital trauma, such as bone fragments within the optic canal, or optic nerve sheath hematomas. Patients with TON usually have severe loss of vision.

pVEP is sensitive to TON, which usually shows significant prolongation of P100 latency and reduction of VEP amplitude [19].

Neurofibromatosis

Neurofibromatosis (NF) is a group of diseases in which tumors form on nerve tissue anywhere in the nervous system. NF is usually diagnosed in childhood or early adulthood. There are three types of NF, including neurofibromatosis type 1 (NF1), neurofibromatosis type 2 (NF2), and schwannomatosis. NF1 and NF2 are related to genetic mutation. Schwannomatosis is related to genetic mutation or unknown causes. In all types of NF, the tumors are usually benign. However, the tumors in some patients can become malignant. NF can cause loss of vision, hearing loss, severe pain, cardiovascular diseases, and impairment of learning.

Electrophysiology can detect the change of visual function in NF. The latency of the P100 peak of pVEP is usually prolonged even without detectable gliomas (Fig. 10.3). If gliomas develop on the optic nerves, then pVEPs can monitor the progression as pressure from the gliomas reduces optic nerve function. The P2 component of fVEP in dark-adapted condition can be absent in a portion of asymptomatic NF patients with normal visual acuity, visual fields, and ophthalmoscopic examination results, which provides a sensitive screening test for asymptomatic patients with NF [20–22]. In NF1 patients, the Arden index of EOG test is significantly higher mainly due to the reduction of dark trough amplitude, which is present in 60% of NF1 patients [23, 24]. The flash ERG does not show significant change in these patients [24].

Optic Nerve Toxicities

Ethambutol

Ethambutol is an oral medication used to treat tuberculosis in combination with other medications. It diffuses into active mycobacterium cells and impairs cellular metabolism, leading to cell death. Optic neuritis is a known adverse effect of this

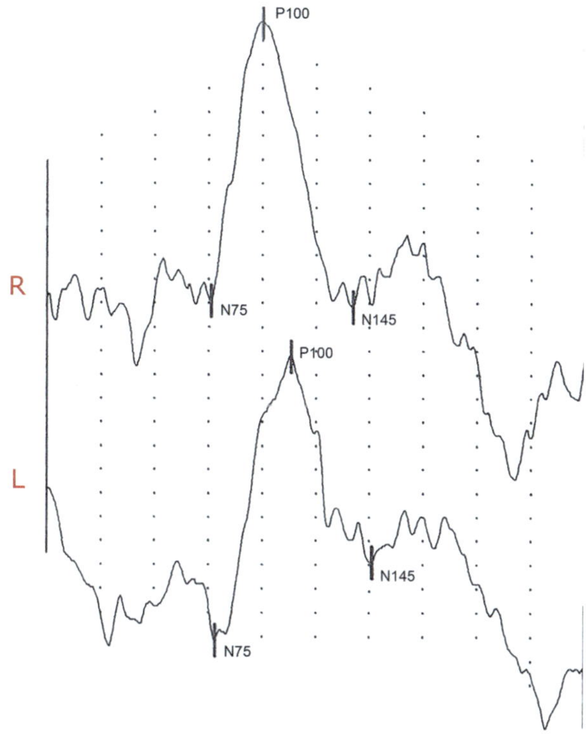

Fig. 10.3 Pattern reversal VEPs recorded from a 5-year-old patient with NF1 showing slowing of VEP components in left optic nerve. Time epoch for VEP is 250 msec

medication and can lead to decrease in visual acuity and irreversible blindness. To prevent damage to the optic nerves during the treatment with ethambutol, it is recommended that visual acuity be tested before and during the therapy. pVEP can also be used to monitor optic nerve function by showing the component P100 delay even while visual acuity, appearance of the ocular fundus, visual fields, color vision, contrast sensitivity, and pupillary light reflex are still normal (Fig. 10.4) [25–27].

Neonatal Hyperbilirubinemia

Neonatal hyperbilirubinemia causes jaundice (a yellow discoloration of the skin and eye's sclera). Jaundice usually becomes visible on sclera at a serum bilirubin level of 2–3 mg/dL, on face at about 4–5 mg/dL, at the umbilicus at about 15 mg/dL, and at the feet at about 20 mg/dL. Hyperbilirubinemia may or may not be harmful depending on its cause and the degree of increase of serum bilirubin level. Some causes of jaundice are harmful no matter what the bilirubin level.

Neurotoxicity is the main concern of neonatal hyperbilirubinemia. fVEP can detect the functional change of the optic nerve. The level of maximal serum bilirubin is positively related to the prolongation of latencies of fVEP [28].

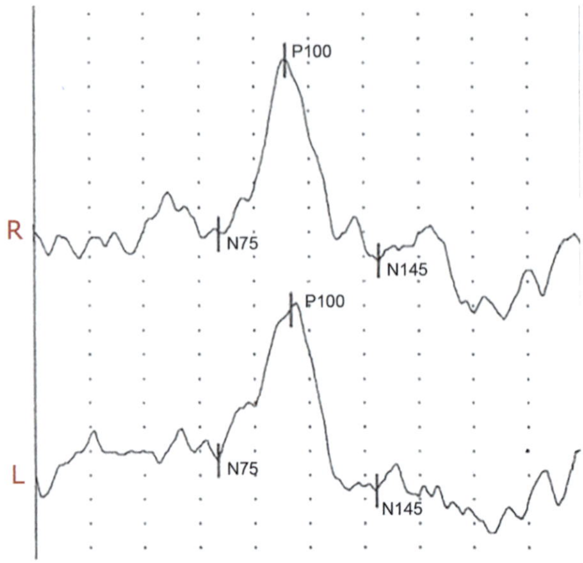

Fig. 10.4 Pattern reversal VEPs recorded from an adult with ethambutol toxicity showing prolonged P100 implicit times of both optic nerve pathways. Time epoch for VEP is 250 msec

Radiotherapy

Head and neck radiotherapy can cause radiation-induced optic neuropathy (RION) [29, 30]. Therefore, detection and monitoring the functional decrease of the optic nerves during radiotherapy for nasopharyngeal carcinoma is necessary to prevent and treat the optic nerve degeneration. pVEP can show prolongation of latency and attenuation of amplitude in VEP in this type of patients many months after radiotherapy (Fig. 10.5) [31].

Leber's Hereditary Optic Neuropathy

Leber's hereditary optic neuropathy (LHON) or Leber's optic atrophy is an inherited disease that causes degeneration of retinal ganglion cells (RGCs) and optic nerve, and loss of central vision. This disease affects young adult males, which often starts as acute and painless loss of central vision. This type of vision loss usually occurs in one eye first. After some weeks, the other eye will have loss of central vision as well. After the acute phase, both eyes continue to have poor central vision. The fundus of LHON can show optic atrophy (Fig. 10.6).

LHON is mainly related to mitochondrial DNA (mtDNA) point mutations [32, 33]. Around 50–70% of LHON cases have G11778A, the position 11778 guanine (G) to adenine (A) mutation. About 30–40% have T14484C, the position 14484

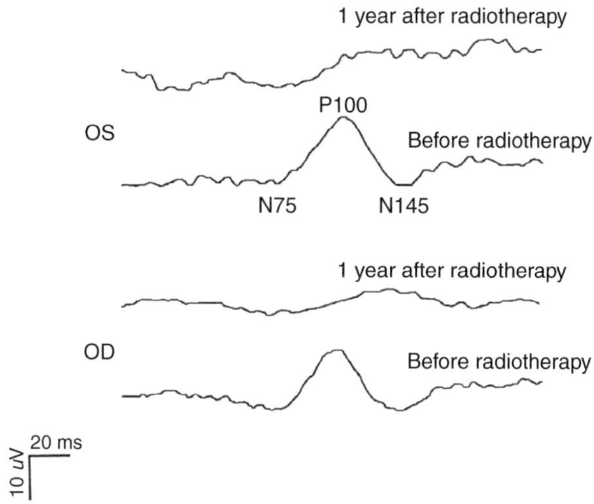

Fig. 10.5 Typical pattern reversal VEP waveforms of a patient before and 1 year after head and neck radiotherapy for nasopharyngeal carcinoma showing the adverse effect on optic nerve pathways (Adapted from Figure 5 [31])

thymine (T) to cytosine (C) mutation or G3460A mutation. Moreover, about 5–10% have other mutations [34, 35]. Because mtDNA mainly exists in egg instead of sperm, LHON mutations may only be passed to the children by the mother and not by the father.

LHON causes optic atrophy. This disease shows prolonged VEP peak times or double-positive peak in VEP waveform after the onset of LHON. With the development of the disease, the VEP amplitude will decrease, and finally the peaks of VEP become indistinguishable (Fig 10.6b) [36, 37]. The amplitude of light-adapted single flash ERG and 30-Hz flicker ERG may also be reduced, implying that the cone photoreceptors are affected by LHON, although there is no change in the appearance of the fundus (Fig 10.6c) [37].

Amblyopia

Amblyopia (lazy eye) is a disease of the visual system due to the abnormal development of retina, optic nerve chiasm, and visual cortex during childhood. Babies are born with visual acuity worse than 20/20 in each eye. Vision improves from birth to about 8 years of age if both eyes are used with the identical focused images projecting onto the retinas. If clear images cannot be received from the retina in one or both eyes, then vision will not develop normally during that period, and the affected eye(s) becomes amblyopic. The abnormal image received by the retina can be caused by strabismus, anisometropia, ametropia, or any disorder that prevents an image to be clearly projected onto the retina (e.g., cataract, clouded cornea, eye bleeding, droopy eyelids).

Amblyopia must be diagnosed and treated as early as possible to avoid permanent vision loss and to allow the development of stereo vision. After about 8 years of age, treatment of amblyopia is unlikely to improve vision.

In all tests of electrophysiology of vision, pVEP is the most sensitive test to detect the functional change in amblyopia from retina to visual cortex, which is usually reflected in latency prolongation and amplitude reduction (Fig. 10.7) [38, 39]. fVEP, used with patients who cannot maintain fixation, might show some change in amblyopia, but it is not as sensitive as pVEP [40]. For the examination of the functional defect in different parts of visual field in amblyopes, multifocal VEP can be used although limited in children because maintaining fixation is required. In 1998, Yu et al., for the first time, explored multifocal VEP in clinical application and found that the multifocal VEP latencies are delayed and amplitudes are reduced in the central/ temporal field in esotropic amblyopia and in central field in anisometropic amblyopia [41], which is consistent with the neural mechanism of suppression in amblyopia [42, 43]. Part of these findings has been confirmed by further studies [44, 45]. pERG can also objectively evaluate the visual function of amblyopic eyes at the level of ganglion cells. ffERGs, focal ERGs, and oscillatory potentials (OPs) are normal in amblyopia [46–48]. The above tests can be selected for differential diagnosis and for the monitoring of functional change in amblyopic eye and the fellow eye during the treatment.

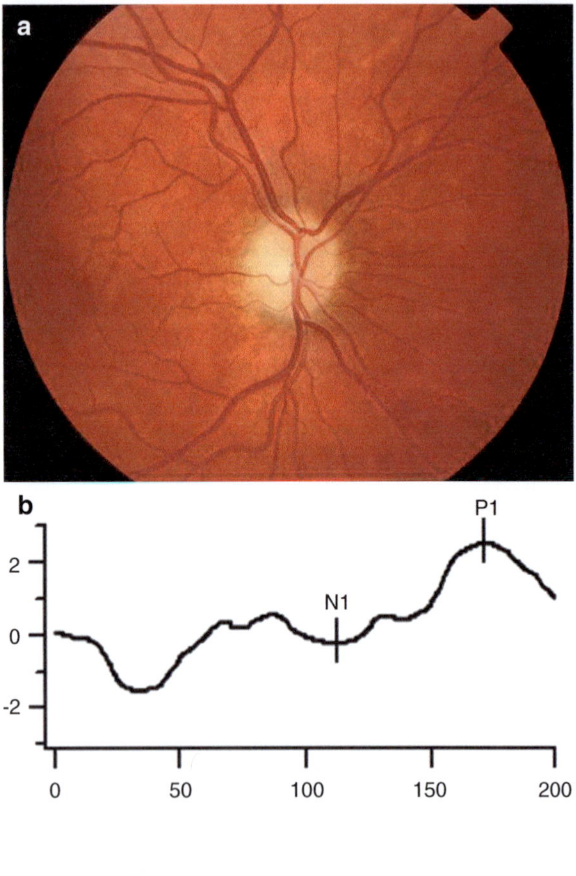

Fig. 10.6 Typical fundus and electrophysiologic waveforms of LHON. (**a**) Representative fundus of male with LHON that shows late stage of optic atrophy. (**b**) VEP waveforms of a LHOH patient. (**c**) ERG waveforms of a LHOH patient. (Figure 10.6b and c are adapted from Figure 3 and 4 [37])

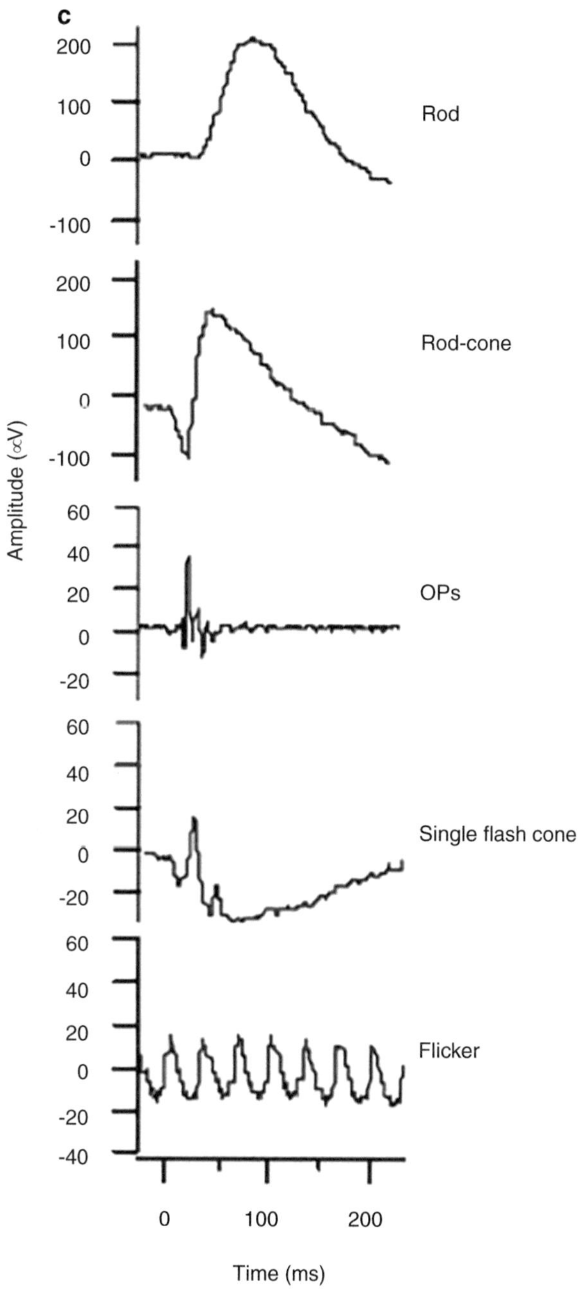

Fig. 10.6 (continued)

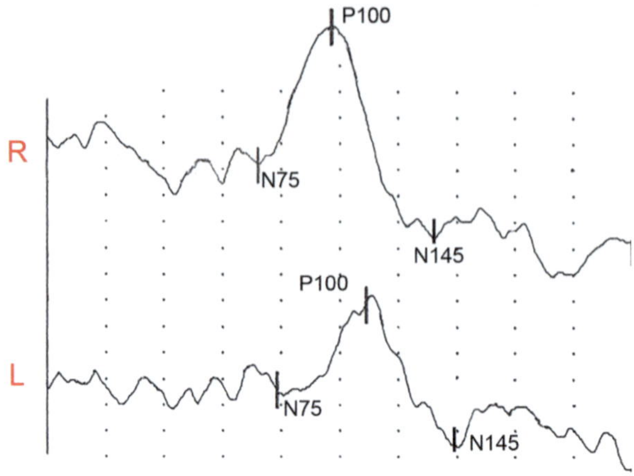

Fig. 10.7 Typical VEP waveforms showing delayed and attenuated P100 in a patient with amblyopia. Time epoch for VEP is 250 msec

References

1. Alshuaib WB. Progression of visual evoked potential abnormalities in multiple sclerosis and optic neuritis. Electromyogr Clin Neurophysiol. 2000;40(4):243–52.
2. Hamurcu M, et al. Analysis of multiple sclerosis patients with electrophysiological and structural tests. Int Ophthalmol. 2017;37(3):649–53.
3. Hayreh SS. Posterior ischaemic optic neuropathy: clinical features, pathogenesis, and management. Eye (Lond). 2004;18(11):1188–206.
4. Hayreh SS. Blood supply of the optic nerve head and its role in optic atrophy, glaucoma, and oedema of the optic disc. Br J Ophthalmol. 1969;53(11):721–48.
5. Isayama Y, et al. Posterior ischemic optic neuropathy. I. Blood supply of the optic nerve. Ophthalmologica. 1983;186(4):197–203.
6. Nuttall GA, et al. Risk factors for ischemic optic neuropathy after cardiopulmonary bypass: a matched case/control study. Anesth Analg. 2001;93(6):1410–6.
7. Ho VT, et al. Ischemic optic neuropathy following spine surgery. J Neurosurg Anesthesiol. 2005;17(1):38–44.
8. Buono LM, Foroozan R. Perioperative posterior ischemic optic neuropathy: review of the literature. Surv Ophthalmol. 2005;50(1):15–26.
9. Lee LA, et al. The American Society of Anesthesiologists Postoperative Visual Loss Registry: analysis of 93 spine surgery cases with postoperative visual loss. Anesthesiology. 2006;105(4):652–9; quiz 867–8.
10. Tormene AP, et al. Electrophysiological findings in anterior ischemic optic neuropathy. Metab Pediatr Syst Ophthalmol (1985). 1989;12(1–3):76–9.
11. Holder GE. Electrophysiological assessment of optic nerve disease. Eye (Lond). 2004;18(11):1133–43.
12. Buono LM, et al. Posterior ischemic optic neuropathy after hemodialysis. Ophthalmology. 2003;110(6):1216–8.
13. Murphy MA. Bilateral posterior ischemic optic neuropathy after lumbar spine surgery. Ophthalmology. 2003;110(7):1454–7.
14. Jayaraman M, et al. Multifocal visual evoked potential in optic neuritis, ischemic optic neuropathy and compressive optic neuropathy. Indian J Ophthalmol. 2014;62(3):299–304.
15. Fraser CL, Holder GE. Electroretinogram findings in unilateral optic neuritis. Doc Ophthalmol. 2011;123(3):173–8.

16. McCulloch DL, et al. Retinal function in infants with optic nerve hypoplasia: electroretinograms to large patterns and photopic flash. Eye (Lond). 2007;21(6):712–20.
17. McCulloch DL, et al. Clinical electrophysiology and visual outcome in optic nerve hypoplasia. Br J Ophthalmol. 2010;94(8):1017–23.
18. Cibis GW, Fitzgerald KM. Optic nerve hypoplasia in association with brain anomalies and an abnormal electroretinogram. Doc Ophthalmol. 1994;86(1):11–22.
19. Ikejiri M, et al. Pattern visual evoked potentials in traumatic optic neuropathy. Ophthalmologica. 2002;216(6):415–9.
20. Jabbari B, et al. The value of visual evoked potential as a screening test in neurofibromatosis. Arch Neurol. 1985;42(11):1072–4.
21. Iannaccone A, et al. Visual evoked potentials in children with neurofibromatosis type 1. Doc Ophthalmol. 2002;105(1):63–81.
22. North K, et al. Optic gliomas in neurofibromatosis type 1: role of visual evoked potentials. Pediatr Neurol. 1994;10(2):117–23.
23. Lubinski W, et al. Supernormal electro-oculograms in patients with neurofibromatosis type 1. Hered Cancer Clin Pract. 2004;2(4):193–6.
24. Lubinski W, et al. Electro-oculogram in patients with neurofibromatosis type 1. Doc Ophthalmol. 2001;103(2):91–103.
25. Kim KL, Park SP. Visual function test for early detection of ethambutol induced ocular toxicity at the subclinical level. Cutan Ocul Toxicol. 2016;35(3):228–32.
26. Yiannikas C, Walsh JC, McLeod JG. Visual evoked potentials in the detection of subclinical optic toxic effects secondary to ethambutol. Arch Neurol. 1983;40(10):645–8.
27. Goyal JL, et al. Evaluation of visual functions in patients on ethambutol therapy for tuberculosis: a prospective study. J Commun Dis. 2003;35(4):230–43.
28. Chen YJ, Kang WM. Effects of bilirubin on visual evoked potentials in term infants. Eur J Pediatr. 1995;154(8):662–6.
29. Mihalcea O, Arnold AC. Side effect of head and neck radiotherapy: optic neuropathy. Oftalmologia. 2008;52(1):36–40.
30. Danesh-Meyer HV. Radiation-induced optic neuropathy. J Clin Neurosci. 2008;15(2):95–100.
31. Hu WH, et al. Impairment of optic path due to radiotherapy for nasopharyngeal carcinoma. Doc Ophthalmol. 2003;107(2):101–10.
32. Wallace DC, et al. Mitochondrial DNA mutation associated with Leber's hereditary optic neuropathy. Science. 1988;242(4884):1427–30.
33. Vergani L, et al. MtDNA mutations associated with Leber's hereditary optic neuropathy: studies on cytoplasmic hybrid (cybrid) cells. Biochem Biophys Res Commun. 1995;210(3):880–8.
34. Soldath P, et al. Leber hereditary optic neuropathy due to a new ND1 mutation. Ophthalmic Genet. 38(5):480–5.
35. Saikia BB, et al. Whole mitochondrial genome analysis in South Indian patients with Leber's hereditary optic neuropathy. Mitochondrion. 2017;36:21–8.
36. Dorfman LJ, et al. Visual evoked potentials in Leber's hereditary optic neuropathy. Ann Neurol. 1977;1(6):565–8.
37. Salomao SR, et al. Visual electrophysiologic findings in patients from an extensive Brazilian family with Leber's hereditary optic neuropathy. Doc Ophthalmol. 2004;108(2):147–55.
38. Arden GB, Barnard WM, Mushin AS. Visually evoked responses in amblyopia. Br J Ophthalmol. 1974;58(3):183–92.
39. Wanger P, Persson HE. Visual evoked responses to pattern-reversal stimulation in childhood amblyopia. Acta Ophthalmol. 1980;58(5):697–706.
40. Davis ET, Bass SJ, Sherman J. Flash visual evoked potential (VEP) in amblyopia and optic nerve disease. Optom Vis Sci. 1995;72(9):612–8.
41. Yu M, Brown B, Edwards MH. Investigation of multifocal visual evoked potential in anisometropic and esotropic amblyopes. Invest Ophthalmol Vis Sci. 1998;39(11):2033–40.
42. Sengpiel F, Blakemore C. The neural basis of suppression and amblyopia in strabismus. Eye (Lond). 1996;10(Pt 2):250–8.

43. Sireteanu R, Fronius M. Naso-temporal asymmetries in human amblyopia consequence of long-term interocular suppression. Vision Res. 1981;21(7):1055–63.
44. Moschos MM, et al. Multifocal visual evoked potentials in amblyopia due to anisometropia. Clin Ophthalmol. 2010;4:849–53.
45. Perez-Rico C, et al. Evaluation of visual function and retinal structure in adult amblyopes. Optom Vis Sci. 2015;92(3):375–83.
46. Arden GB, Wooding SL. Pattern ERG in amblyopia. Invest Ophthalmol Vis Sci. 1985;26(1):88–96.
47. Ikeda H, Tremain KE. Amblyopia occurs in retinal ganglion cells in cats reared with convergent squint without alternating fixation. Exp Brain Res. 1979;35(3):559–82.
48. Wanger P, Persson HE. Oscillatory potentials, flash and pattern-reversal electroretinograms in amblyopia. Acta Ophthalmol. 1984;62(4):643–50.

Chapter 11
Characteristics of Visual Electrophysiology in Albinism

Donnell Creel, Minzhong Yu, and Alessandro Iannaccone

A feature of the mammalian visual system is the development of optic fibers originating in temporal retina that do not cross at the optic chiasm. Most vertebrates originating prior to mammals have complete crossing of optic fibers at the chiasm. In mammals, eyes move forward on the head from those species with laterally placed eyes, such as in guinea pigs, through rodents and lagomorphs, horses, dogs, and cats as examples, reaching maximum frontal placement in primates. With this progression is an increase in number of optic nerve fibers remaining uncrossed at the chiasm as the proportion of temporal retina increases, along with binocular overlap of the visual fields. In humans, the proportion of optic nerve retinal ganglion cells not crossing at the chiasm is near 45%.

Around 90–99% of rodent and lagomorph retinal ganglion cells cross at the optic chiasm (Fig. 11.1). When terminating in the geniculate nuclei, they do not exhibit the laminated dorsal lateral geniculate nucleus (dLGN) seen in carnivores and primates including human beings. Carnivore and primates' dorsal lateral geniculate nuclei (dLGN) contain up to six alternating monocular layers with point-to-point representation of visual space passed on to the primary visual cortex. The 90%+ crossed RGCs at the optic chiasm in rodents and lagomorphs fill the contralateral dLGN, with only a pocket of the ipsilateral dLGN devoted to the fewer uncrossed fibers. The small number of uncrossed fibers in albino rodents project to the dLGN in a fragmented fashion as opposed to a well-organized pattern in pigmented rodents [1].

D. Creel (✉)
Moran Eye Center, University of Utah School of Medicine, Salt Lake City, UT, USA
e-mail: donnell.creel@hsc.utah.edu

M. Yu
Department of Ophthalmology, University Hospitals Eye Institute, Cleveland, OH, USA

A. Iannaccone
Center for Retinal Degenerations and Ophthalmic Genetic Diseases, Duke University School of Medicine, Duke Eye Center, Department of Ophthalmology, Durham, NC, USA

M. Yu et al. (eds.), *Handbook of Clinical Electrophysiology of Vision*,
https://doi.org/10.1007/978-3-030-30417-1_11

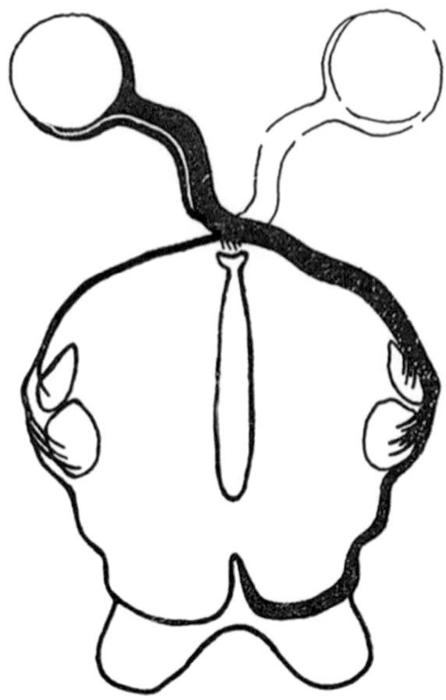

Fig. 11.1 Schematic of range of crossed and uncrossed optic projections at the chiasm in rodents

The organization of the path retinal ganglion cells take at the chiasm also changes as the uncrossed optic system progresses up the phylogenetic scale from rodents to primates. In rodents, the uncrossed optic fibers haphazardly start the older phylogenetic route to the center of the chiasm, and then a few appear to bounce back coursing to the ipsilateral side of the brain (Fig. 11.2). As the organization of the non-crossing optic system evolves to primates, the non-crossing fibers become an organized bundle in the lateral portion of the optic chiasm [2].

Until 1965, no one suggested there might be dramatic difference in the visual systems within a species. After comparing the interocular visual pathways in "split brain" hooded rats with ocular pigmentation and albino rats, Sheridan concluded that "Perhaps the paucity of uncrossed fibers that characterized rodents in general is even further reduced in the albino" [3]. Lund verified Sheridan's hypothesis anatomically. Lund stated albino rats display no organized uncrossed optic fibers [4].

For several years, the anatomical anomaly of few uncrossed retinal ganglion cells (RGCs) at the optic chiasm in albinos appeared to be limited to rats and rabbits [5]. In 1969, Guillery described aberrant retinogeniculate organization in Siamese cats, but the connection to albinism was not recognized [6]. The first link between Siamese cats and albinism, and the hypothesis that reduced uncrossed optic fibers is likely a "highly general transspecies phenomenon in albinic mammals" was suggested by Creel in 1971 [7, 8]. Siamese cats possess a tyrosinase-locus mutation that is a temperature-sensitive pigmentation defect (i.e., pigment forms only on the cold parts of the body similar to Himalayan mice, rats, and rabbits). This mutation is referred to as the Siamese–Himalayan temperature effect. The best figures showing the consequences of reduced uncrossed retinal ganglion cells in Siamese cats includ-

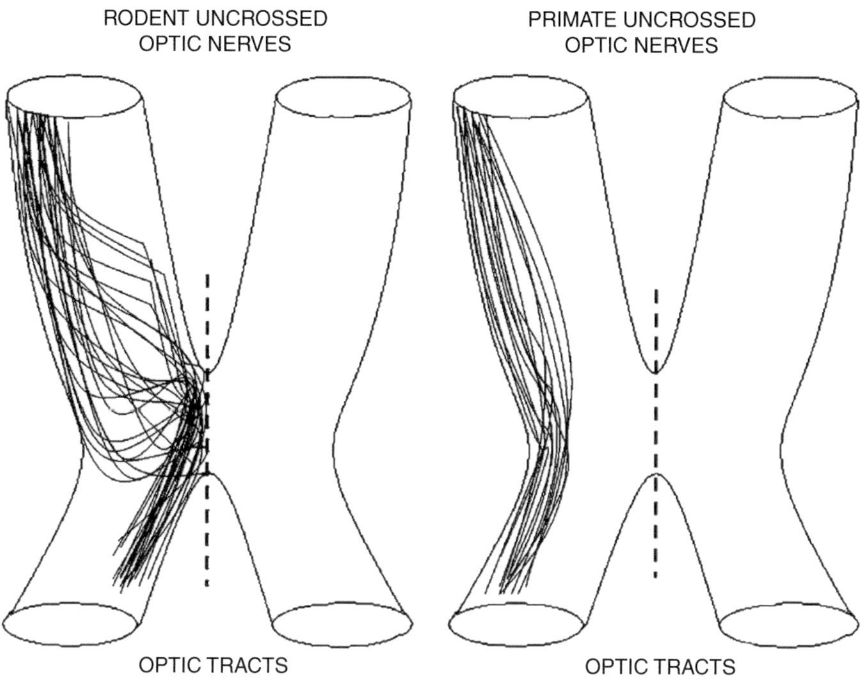

Fig. 11.2 In rodents, the uncrossed optic fibers are disorganized starting the older phylogenetic route to the center of the chiasm mixing with RGCs from other eye and then reverting to the ipsilateral side of the brain. In primates, the non-crossing fibers are an organized bundle in the lateral portion of the optic chiasm (Adapted from Neuveu & Jeffery. Figure owned by Springer)

ing lateral geniculate to visual cortex are in Guillery's 1974 review article in Scientific American [9].

All mammals studied that lack embryonic retinal melanin pigment in the developing retina have excessive crossing of retinogeniculate optic fibers at the chiasm. This misrouting is associated with a number of visual system abnormalities that are most apparent in primates including humans: the fovea of the retina is not normally developed, vascular microcapillaries intrude into foveal avascular zone, the brainstem centers controlling eye movement receive abnormal input from the retina, the laminations of the dorsal lateral geniculate are fragmented, and at the cortical level, the connections for stereovision are sparse. These anomalies appear in all mammals with insufficient embryonic retinal melanin pigment in the pigment epithelium from mouse to humans. Only ocular albinism (OA) is required to reduce uncrossed optic nerve fibers at the chiasm.

The two most common types of albinism are Type 1 associated with errors in oxidation of tyrosine and DOPA located on chromosome 11q 14–21 and Type 2 associated with errors in melanosomal membrane protein metabolism. This is the most common gene associated with albinism in sub-Saharan Africa located long arm 15q 11–12. Nearby genes on chromosome 15q extending from 15q 13–21, including Prader–Willi syndrome, Angelman syndrome, and possibly some with Asperger syndrome, have been found to have optic misrouting.

Classification of forms of albinism is based on genetic locus. Chromosomal location has been identified for over 20 forms of human albinism on 12 different chromosomes: oculocutaneous albinism (OCA1–7), ocular albinism (OA1), 10 forms of Hermansky–Pudlak syndrome (HPS1–10) [10], Chediak–Higashi syndrome (CHS), and Griscelli syndrome (GS) [11, 12]. An additional syndromic form of albinism added is the rare Vici syndrome (VICIS) [13, 14]. Phenotypic expression varies considerably within genetic forms of albinism and even between siblings [15, 16].

Visual abnormalities described in albinism do not include disorders of hypopigmentation due to embryonic migration defects of pigment cells from the neural crest such as in Waardenburg syndrome and deaf white cat (W gene), vitiligo or piebald spotting, nor other forms of hypopigmentation such as phenylketonuria. These hypopigmenting conditions do not show the visual anomalies.

Individuals with chiasmic misrouting are not required to be classic forms of albinism. Currently, over 20 genetic forms of albinism in humans are recognized, with more likely to be discovered, because more have been identified in mouse models. Misrouted optic fibers at the chiasm are constant in all forms of retinal hypopigmentation. Individuals with these visual abnormalities, particularly those associated with rare syndromes such as the Vici Syndrome, are likely underdiagnosed. Occasionally, patients with misrouted optic fibers are likely due to linkage between nearby albinism genes on same chromosome (e.g., Prader–Willi syndrome) [17]. Recording monocular VEPs across occipital cortex is the easiest albino concomitant to measure. Albino genetic panels are expensive and only available at a couple of institutions.

In normally pigmented individuals, the ganglion cell axons originating from the nasal side of the fovea cross to the contralateral side at the optic chiasm, and those from temporal side of the fovea stay on the same side at the optic chiasm. However, in humans with hypopigmented retinas, almost all retinal ganglion cells originating from up to 15° in temporal retina also cross to the contralateral side at the optic chiasm (Fig. 11.3). The exact proportion of these ganglion cells in humans is not known, which likely varies considerably due to amount of residual pigment plus individual genetic differences including number of loci with mutations.

In many albino animals, essentially all optic neurons from nasal and temporal retina cross at the optic chiasm. As shown in Fig. 11.3, it is estimated that 80% of the ganglion cells cross to contralateral side with a 15-degree shift of vertical meridian to temporal side in human albinos, based on extrapolating from animal anatomical studies and multifocal visually evoked potential studies [18].

Ocular albinism (OA) and oculocutaneous albinism (OCA) are characterized by photophobia, nystagmus, and reduced visual acuity. On anterior segment examination, there are often iris transillumination defects, and on fundus examination, there is foveal hypoplasia and a hypopigmented fundus (Fig. 11.4). While OCA has autosomal recessive inheritance, ocular albinism has both X-linked (XLOA) and recessive forms (OAR) [19, 20]. As indicated by the name, OCA causes skin

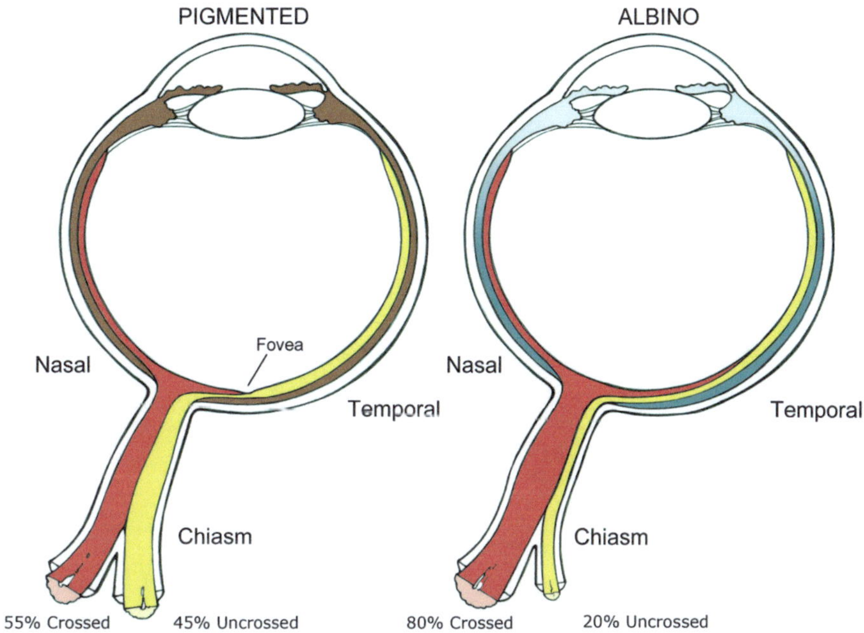

Fig. 11.3 Diagram of approximate distribution of the optic nerve ganglion cells at the optic chiasm in normal and albino human beings

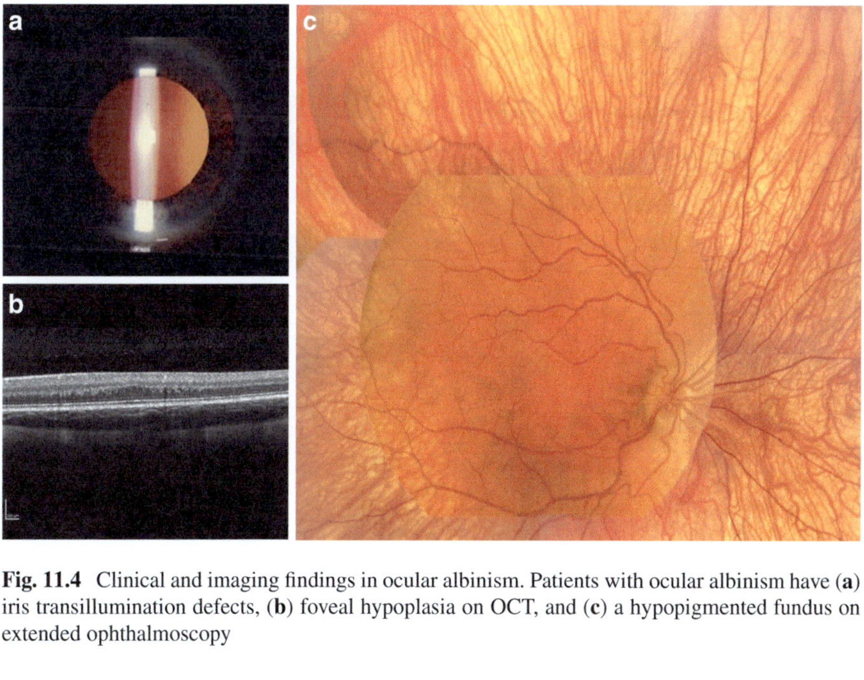

Fig. 11.4 Clinical and imaging findings in ocular albinism. Patients with ocular albinism have (**a**) iris transillumination defects, (**b**) foveal hypoplasia on OCT, and (**c**) a hypopigmented fundus on extended ophthalmoscopy

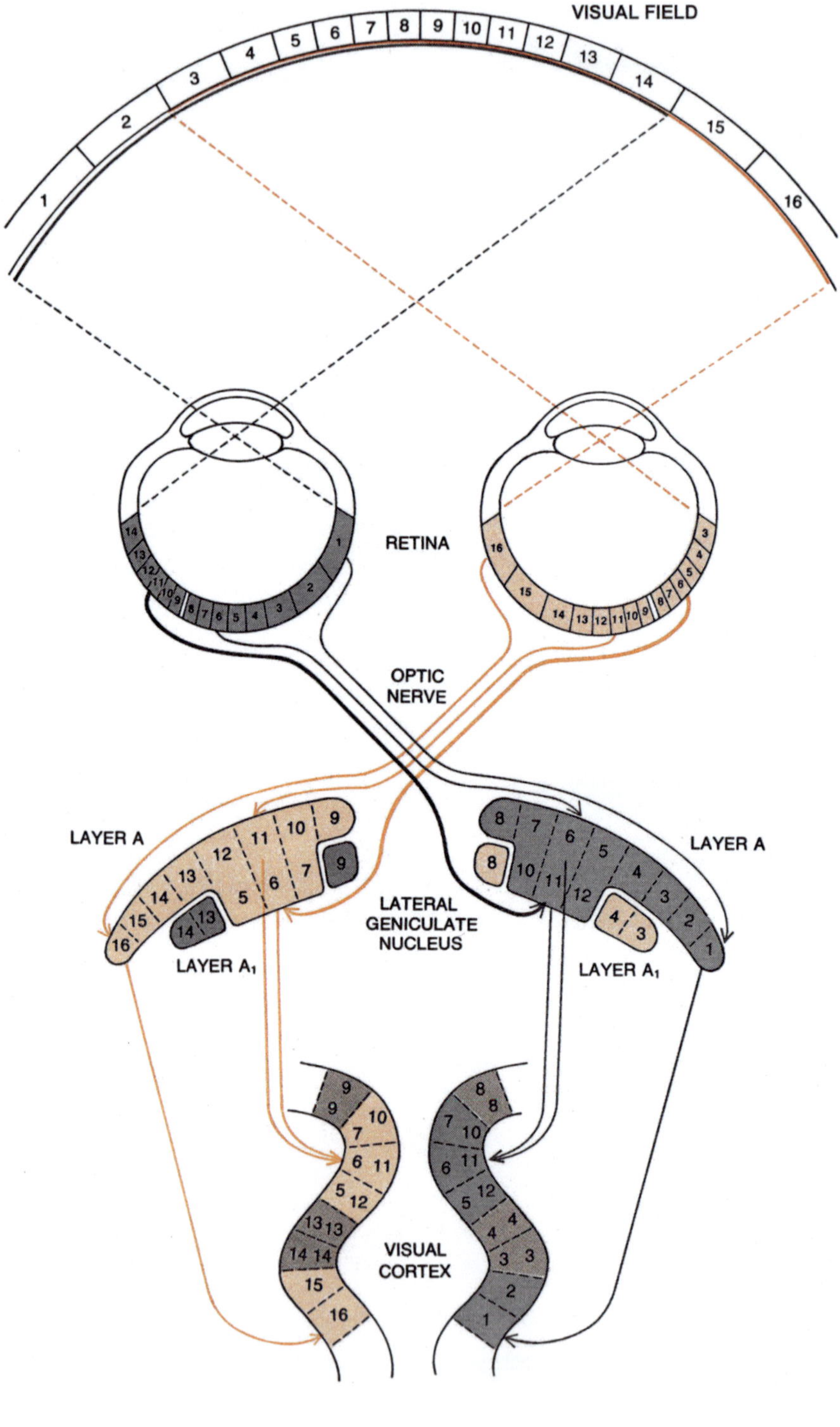
VISUAL FIELD
RETINA
OPTIC
NERVE
LAYER A
LAYER A
LATERAL
GENICULATE
NUCLEUS
LAYER A₁
LAYER A₁
VISUAL
CORTEX

Fig. 11.5 Disorganization of central visual pathways in albinism. Schematic example of disruption and fragmentation of the lamination of LGNd in Siamese cat producing serious misrepresentation of the visual field to cortex reducing the number of cortical binocular cells. Figure from R.W. Guillery article in Scientific American, May, 1974. We thank the estate of the artist Bunji Tagawa for permission to use this figure

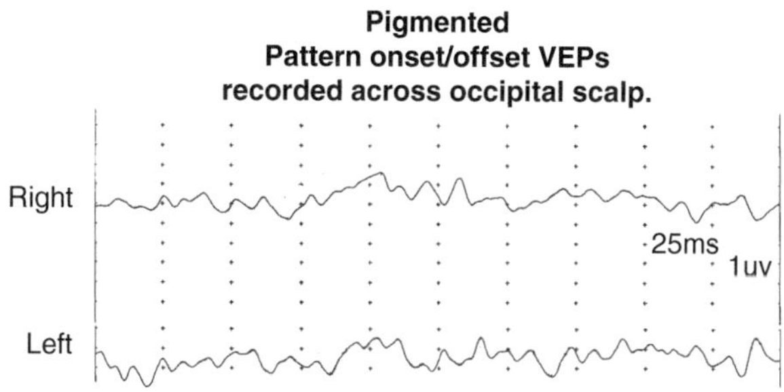

Fig. 11.6 Typical monocular VEP waveforms recorded across occipital scalp of a normally pigmented individual showing little change in waveform when each eye is stimulated monocularly. Time epoch 250 msec

hypopigmentation in addition to ocular manifestations. Skin pigmentation is usually normal in all forms of ocular albinism, but unlike OAR, OA has macromelanosomes in both skin and retinal pigment epithelium (RPE) [21].

In XLOA, OAR, and OCA, ffERGs are typically normal to supernormal due to an eye with less pigment. Thus, since achromatopsia (ACHM) and blue cone monochromacy (BC) are part of the differential diagnostic process in patients affected by OA, the ffERG is essential in distinguishing OA from the former conditions. The only consistent and typical visual electrophysiological abnormality in OA and OCA patients is an asymmetric cortical VEP response distribution due to the abnormal visual pathway decussation at the chiasm that characterizes all these conditions [22]. Cats and primates have distinct segregation of crossed and uncrossed retinal ganglion cell terminating in layers of dorsal lateral geniculate nuclei (LGNd). The consequences of retinal ganglion fibers from temporal retina also crossing at the chiasm is they encroach into and fragment the adjacent layer. In normally pigmented cats and primates, the points in the visual field are maintained in columns through the layers of crossed and uncrossed ganglion cell terminations, such as points 13 and 14 in Fig. 11.5. Ganglion cell terminations originating from temporal retina crossing abnormally invade the adjacent layer, such as points 5, 6 and 7 that is carried on to cortex disrupting the anatomical basis for cortical binocular stereovision.

In 1974, using scalp-recorded visually evoked potentials (VEPs), Creel et al. reported four genetic forms of human oculocutaneous albinism exhibited electrophysiologic evidence of reduced uncrossed optic fibers [22]. Similar results were published for human ocular albinos [23] and replicated in early studies by Taylor (1978) [24] and Coleman et al. (1979) [25]. Reviews of research in this area include Guillery (1974, 1996) [26], Hoffmann et al. (2006) [18], Neveu and Jeffery (2007) [2], and Bridge et al. (2014) [27].

The best visual stimulus for recording VEPs from albinos is a pattern onset/offset with individual checks subtending 1° of arc or larger. For individuals with very

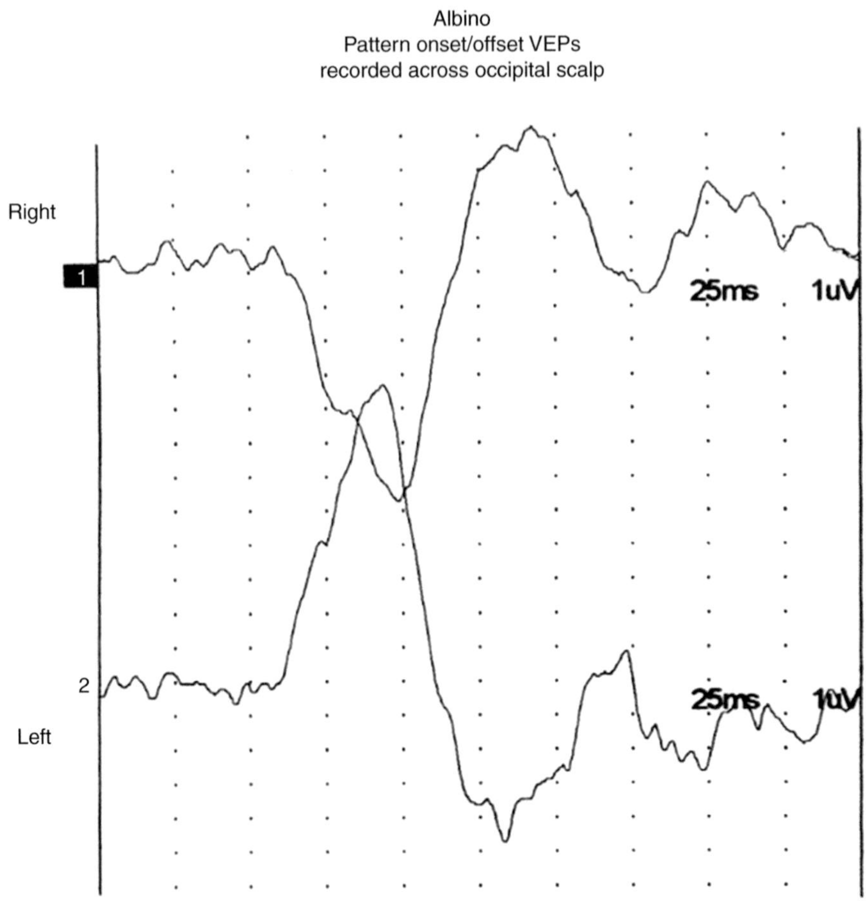

Fig. 11.7 Typical monocular VEP waveforms recorded across occipital scalp of an albino showing reversal of polarity when each eye is stimulated. This reflects that almost all optic nerve fibers cross at the optic chiasm to opposite hemisphere in albino mammals. Time epoch for VEP 250 msec

poor vision and infants, flash stimuli are preferred. In some albinos, comparing evoked potentials recorded from the International 10–20 System locations O1 and O2 [28] demonstrate the misrouting. In albinos, other more lateral scalp locations along the same Oz plane about halfway to T5 and T6 (i.e., H3 and H4), often better demonstrate misrouting. An early finding was that, with monocular stimulation, the "difference" potential between 10 and 20 International System scalp locations O1–O2 or H3–H4 across the midline was an efficient way to exhibit chiasmic misrouting in albinos [29]. The most discriminating VEP components are the found in the 50–200 millisecond time window of pattern onset or flash VEP. The worst stimulus is pattern reversal that exacerbates nystagmus in albinos and consequently blurs the image.

To detect misrouting at the chiasm, the following two recording sites are required: one positive electrode and one negative location in each occipital hemisphere about 3 cm above the inion and 4 cm or more off the midline in an adult. This bipolar montage records the "difference" potential between the right and left hemispheres, requiring no additional reference location. The ground electrode can be attached anywhere.

In normal subjects with pigmented retinae, innervation of visual cortex from nasal and temporal retinas is comparable, so the "difference" potential recorded between hemispheres would be nearly zero or appear similar no matter which eye is stimulated. In most cases, the recorded VEP in normal subjects is not completely flat because the projection of visual fields on occipital poles is usually a little asymmetrical displayed in the VEPs in Fig. 11.6.

In contrast, the majority of central retinal RGC fibers cross to the contralateral hemisphere in albinos so there is a marked "difference" potential recorded between the two occipital hemispheres showing opposite polarity of the VEP waveforms across the occipital area between stimulating the right eye and the left eye (Fig. 11.7).

Mechanisms affecting visual system development are well known. See Jeffery and Erskine (2005) and Herrera et al. (2019) for theories of molecular mechanisms directing ganglion cell routing at chiasm. Many candidate agents, including signaling and transcription factors, have been suggested to be associated with retinal ganglion cells coursing through the chiasm and other visual features in albinism. Likely many of these conclusions are correct within the model studied, but not a general solution within mammals. Sprinkled through publications since the 1970s are suggestions that genetic effects are paramount. A genetic argument suggested by Creel [30] proposes that the lack of melanin pigment initiates atavistic expression of visual and auditory embryogenesis. Melanin pigments in the nascent retina, or likely genetic coding for melanin pigment, are the positive inductive signal launching normal retinal development and retinal ganglion cell targeting. When retinal melanin is reduced, retinal embryogenesis defaults to the conserved genetic package existing when the species first evolved with a cascade of consequences.

The vertebrate genome includes coding of genetic past. Mammals retain the evolutionary base instruction package for complete crossing of the retinal ganglion cells in ancestral vertebrates, and little genetic information concerning the development of area centralis, fovea, or vascular sparing of the foveal area. The essence of this theory is that mammals with nondecussating retinal ganglion cells are genetically unstable. Retinal melanin pigment is so fundamental to retinal vascular cells and early retinal ganglion cells that insufficient retinal melanin triggers a switch resulting in the phylogenetically new genetic instructions not to be completed. In albinos, genetic instruction defaults to the simpler and more entrenched platform. Phylogenetically late genetic addendums to conserved instructions are vulnerable to not being expressed if genetic cues, such as sufficient retinal melanin pigmentation, are not present to support variation from highly conserved coding.

References

1. Creel D, Giolli RA. Retinogeniculate projections in albino and ocularly hypopigmented rats. J Comp Neurol. 1976;166(4):445–55.
2. Neveu MM, Jeffery G. Chiasm formation in man is fundamentally different from that in the mouse. Eye (Lond). 2007;21(10):1264–70.
3. Sheridan CL. Interocular transfer of brightness and pattern discriminations in normal and corpus callosum-sectioned rats. J Comp Physiol Psychol. 1965;59:292–4.
4. Lund RD. Uncrossed visual pathways of hooded and albino rats. Science. 1965;149(3691):1506–7.
5. Giolli RA, Guthrie MD. The primary optic projections in the rabbit. An experimental degeneration study. J Comp Neurol. 1969;136(1):99–126.
6. Guillery RW. An abnormal retinogeniculate projection in Siamese cats. Brain Res. 1969;14(3):739–41.
7. Creel DJ. Visual system anomaly associated with albinism in the cat. Nature. 1971;231(5303):465–6.
8. Creel DJ. Differences of ipsilateral and contralateral visually evoked responses in the cat: strains compared. J Comp Physiol Psychol. 1971;77(1):161–5.
9. Guillery RW. Visual pathways in albinos. Sci Am. 1974;230(5):44–54.
10. Summers CG, Hand JL, Hermansky-Pudlak syndrome. 2019. https://www.uptodate.com/contents/hermansky-pudlak-syndrome.
11. Dotta L, et al. Clinical, laboratory and molecular signs of immunodeficiency in patients with partial oculo-cutaneous albinism. Orphanet J Rare Dis. 2013;8:168.
12. Montoliu L, et al. Increasing the complexity: new genes and new types of albinism. Pigment Cell Melanoma Res. 2014;27(1):11–8.
13. Dionisi Vici C, et al. Agenesis of the corpus callosum, combined immunodeficiency, bilateral cataract, and hypopigmentation in two brothers. Am J Med Genet. 1988;29(1):1–8.
14. Filloux FM, et al. Ophthalmologic features of Vici syndrome. J Pediatr Ophthalmol Strabismus. 2014;51(4):214–20.
15. Summers CG, Hand JL, Oculocutaneous albinism. 2019. https://www.uptodate.com/contents/oculocutaneous-albinism.
16. Cheong PY, King RA, Bateman JB. Oculocutaneous albinism: variable expressivity of nystagmus in a sibship. J Pediatr Ophthalmol Strabismus. 1992;29(3):185–8.
17. Creel DJ, et al. Abnormalities of the central visual pathways in Prader-Willi syndrome associated with hypopigmentation. N Engl J Med. 1986;314(25):1606–9.
18. Hoffmann MB, et al. Assessment of cortical visual field representations with multifocal VEPs in control subjects, patients with albinism, and female carriers of ocular albinism. Invest Ophthalmol Vis Sci. 2006;47(7):3195–201.
19. Hutton SM, Spritz RA. A comprehensive genetic study of autosomal recessive ocular albinism in Caucasian patients. Invest Ophthalmol Vis Sci. 2008;49(3):868–72.
20. Kamaraj B, Purohit R. Mutational analysis of oculocutaneous albinism: a compact review. Biomed Res Int. 2014;2014:905472.
21. Iannaccone A, et al. Identification of two novel mutations in families with X-linked ocular albinism. Mol Vis. 2007;13:1856–61.
22. Creel D, Witkop CJ Jr, King RA. Asymmetric visually evoked potentials in human albinos: evidence for visual system anomalies. Investig Ophthalmol. 1974;13(6):430–40.
23. Creel D, O'Donnell FE Jr, Witkop CJ Jr. Visual system anomalies in human ocular albinos. Science. 1978;201(4359):931–3.
24. Taylor WO. Edridge-Green Lecture, 1978. Visual disabilities of oculocutaneous albinism and their alleviation. Trans Ophthalmol Soc U K. 1978;98(4):423–45.
25. Coleman J, et al. Abnormal visual pathways in human albinos studied with visually evoked potentials. Exp Neurol. 1979;65(3):667–79.

26. Guillery RW. Why do albinos and other hypopigmented mutants lack normal binocular vision, and what else is abnormal in their central visual pathways? Eye (Lond). 1996;10(Pt 2):217–21.
27. Bridge H, et al. Changes in brain morphology in albinism reflect reduced visual acuity. Cortex. 2014;56:64–72.
28. Jasper HH. Report of committee on methods of clinical examination in electroencephalography. Electroencephalogr Clin Neurophysiol. 1958;10:370–5.
29. Creel D, Spekreijse H, Reits D. Evoked potentials in albinos: efficacy of pattern stimuli in detecting misrouted optic fibers. Electroencephalogr Clin Neurophysiol. 1981;52(6):595–603.
30. The Organization of the Retina and Visual System. Webvision. Moran Eye Center, 2014. Web: https://webvision.med.utah.edu/book/electrophysiology/visual-and-auditory-anomalies-associated-with-albinism/

Index

M. Yu et al. (eds.), *Handbook of Clinical Electrophysiology of Vision*,
https://doi.org/10.1007/978-3-030-30417-1

MIX
Papier aus verantwortungsvollen Quellen
Paper from responsible sources
FSC® C105338

If you have any concerns about our products,
you can contact us on
ProductSafety@springernature.com

In case Publisher is established outside the EU,
the EU authorized representative is:
Springer Nature Customer Service Center GmbH
Europaplatz 3, 69115 Heidelberg, Germany

Printed by Libri Plureos GmbH
in Hamburg, Germany